·四川大学精品立项教材·

Advanced Calculation Mechanics

高等计算力学

Zheming Zhu Xingyu Wang Li Ren Meng Wang
朱哲明　王兴渝　任　利　王　蒙　编著

四川大学出版社
Sichuan University Press

责任编辑：毕　潜
责任校对：蒋　玙
封面设计：墨创文化
责任印制：王　炜

图书在版编目(CIP)数据

高等计算力学：英文／朱哲明等编著．—成都：四川大学出版社，2019.3
ISBN 978-7-5690-2830-0

Ⅰ.①高…　Ⅱ.①朱…　Ⅲ.①计算力学-英文
Ⅳ.①O302

中国版本图书馆 CIP 数据核字（2019）第 049916 号

书名　**高等计算力学**
Advanced Calculation Mechanics

编　著　朱哲明　王兴渝　任　利　王　蒙
出　版　四川大学出版社
地　址　成都市一环路南一段 24 号 (610065)
发　行　四川大学出版社
书　号　ISBN 978-7-5690-2830-0
印　刷　郫县犀浦印刷厂
成品尺寸　185 mm×260 mm
印　张　12.75
字　数　416 千字
版　次　2019 年 3 月第 1 版
印　次　2019 年 3 月第 1 次印刷
定　价　48.00 元

◆读者邮购本书，请与本社发行科联系。
电话：(028)85408408/(028)85401670/
(028)85408023　邮政编码：610065
◆本社图书如有印装质量问题，请寄回出版社调换。
◆网址：http://press.scu.edu.cn

Preface

The advent of high-speed computers has given tremendous impetus to all numerical methods for solving engineering problems. Finite element method forms one of the most versatile classes of such methods, and was originally developed in the field of structural analysis. Finite element method is a computer-aided mathematical tool which can be used for obtaining approximate solutions of some parameters to those engineering problems that can be represented by physical system subjected to external influences. Such problems are in areas like solid mechanics (elasticity, plasticity, statics, dynamics, etc.), heat transfer (conduction, convection, radiation), fluid mechanics, electromagnetism and coupled interactions of the above, e. g. fluid-solid interaction. Application in solid mechanics is much more extensive, and can be classified in many different ways depending on the application types, physical system shapes or responses, loading conditions, and so on.

The purpose of this course is to introduce students (both undergraduate and graduate students) to the fundamentals of the finite element method and its applications in engineering with emphasis on solid structures or liquid. The students learn the necessary principle to help them create a basic finite element code (using FORTRAN or MATLAB or C programming language) for the stress and deformation analysis of solid structures.

This book is based on courses given by the author to both undergraduate and graduate students of engineering mechanics at Sichuan University. A prior knowledge of the FORTRAN or MATLAB or C programming language and solid mechanics is assumed. The level of continuum mechanics, numerical analysis, matrix algebra and other mathematics employed in this book is that normally taught in undergraduate engineering courses. The book is therefore suitable for engineering undergraduates and other students at an equivalent level. Postgraduates and practising engineers may also find it useful if they are comparatively new to finite element methods.

Sichuan University, Zheming Zhu

Chengdu, China

Contents

Chapter 1 Discretization and element stiffness

The limitations of the human mind are such that it cannot grasp the behaviour of its complex surroundings and creations in one operation. Therefore, the process of subdividing all systems into their individual components or 'elements', whose behavior is readily understood, and then rebuilding the original system from such components to study its behavior, is a natural way in which the engineer, the scientist, or even the economist proceeds.

In many situations an adequate model is obtained using a finite number of well-defined components, and we shall term such problems discrete. In others the subdivision is continued indefinitely and the problem can only be defined using the mathematical fiction of an infinitesimal. This leads to differential equations or equivalent statements which imply an infinite number of elements, and we shall term such systems continuous.

With the advent of digital computers, discrete problems can generally be solved readily even if the number of elements is very large. As the capacity of all computers is finite, continuous problems can only be solved exactly by mathematical manipulation. Here, the available mathematical techniques usually limit the possibilities to oversimplified situations.

To overcome the intractability of realistic types of continuum problems, various methods of discretization have been proposed from time to time both by engineers and mathematicians. All involve an approximation which, hopefully, approaches in the limit the true continuum solution as the number of discrete variables increases.

The discretization of continuous problems has been approached differently by mathematicians and engineers. Mathematicians have developed general techniques applicable directly to differential equations governing the problem, such as finite difference approximations, various weighted residual procedures, or proximate techniques for determining the stationarity of properly defined 'functionals'. The engineer, on the other hand, often approaches the problem more intuitively by creating an analogy between real discrete elements and finite portions of a continuum domain.

Since the early 1960s much progress has been made, and today the purely mathematical and 'analogy' approaches are fully reconciled. It is the object of this text to present a view of the finite element method as a general discretization procedure of continuum problems posed by mathematically defined statements.

In the analysis of problems of a discrete nature, a standard methodology has been developed over the years. The civil engineer, dealing with structures, first calculates force-displacement relationships for each element of the structure and then proceeds to assemble the whole by following a well-defined procedure of establishing local equilibrium at each 'node' or connecting point of the structure. The resulting equations can be solved for the unknown displacements. Similarly, the electrical or hydraulic engineer, dealing with a network of electrical components (resistors, capacitances, etc.) or hydraulic conduits, first establishes a relationship between currents (flows) and potentials for individual elements and then proceeds to assemble the system by ensuring continuity of flows.

All such analyses follow a standard pattern which is universally adaptable to discrete systems. It is thus possible to define a standard discrete system, and this chapter will be primarily concerned with establishing the processes applicable to such systems. Much of what is presented here will be known to engineers, but some reiteration at this stage is advisable. As the treatment of elastic solid structures has been the most developed area of activity, this will be the focus of this book.

1.1 Discretization of a domain by elements

Forstructures with a complex configuration and subjected to multi-loads as shown in Fig. 1-1, it is very difficult, and sometimes it is impossible, to obtain analytical solutions of the stresses or the displacements by using the knowledge of solid mechanics. Under such scenario, a better approach is to discretize the structure or the domain into a finite number of small triangular (or quadrilateral) elements, as shown in Fig. 1-2, and then try to solve the stresses and/or strains of each element. This process is called discretization of a domain. For three dimensional problems, we use tetrahedral elements or brick elements to discretize the domain.

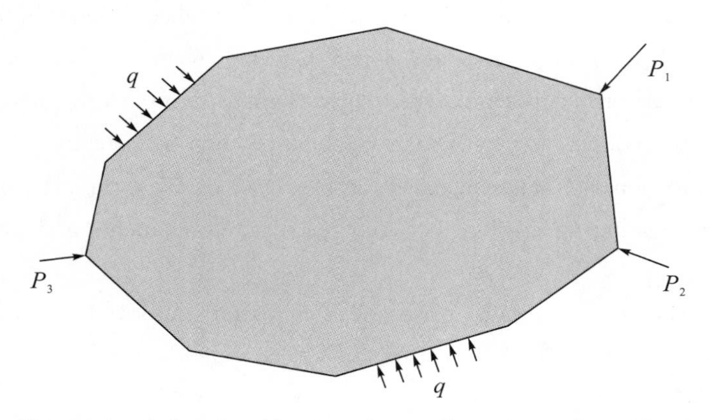

Fig. 1-1 A domain with a complex configuration and multi-loads

Assume that:

1. The continuum (or domain) is separated by imaginary lines or surfaces (in 3D cases) into a number of 'finite elements', and the nodes are numbered. The number of the nodes are arbitrary, but for saving the calculation time, the difference between a node number with its adjacent node numbers should be as small as possible.
2. The elements are interconnected at a discrete number of nodes, not the imaginary lines. If there is action between two adjacent elements, only the conjunct nodes can transfer the force or the displacement.
3. The phenomena of overlap and gap between two adjacent elements as shown in Fig. 1−2 are not allowed.

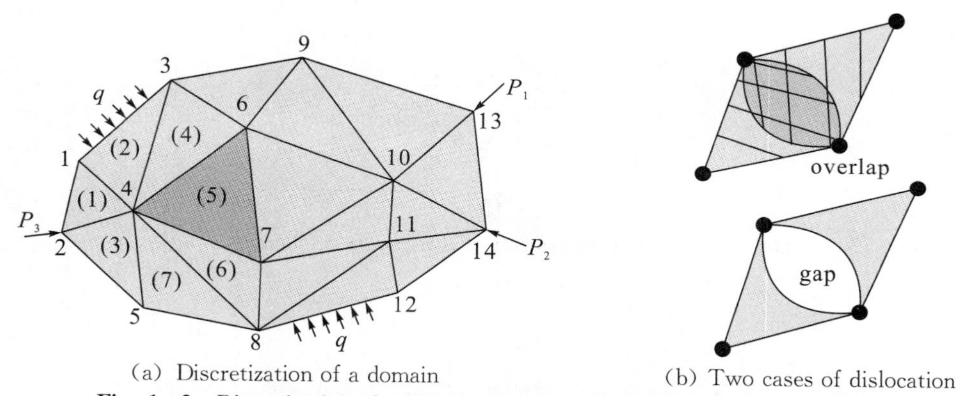

(a) Discretization of a domain (b) Two cases of dislocation

Fig. 1−2 Discretization of a domain with triangular elements and numbered nodes

After the domain is discretized by a finite number of triangle elements, as shown in Fig. 1−3, we will try to solve the stresses and/or the strains of each element. First, we arbitrarily take one triangular element from the domain divided by the triangular elements, for example, the element number (5), and for generality, the node numbers are replaced by $i\,(x_i,\ y_i)$, $j\,(x_j,\ y_j)$ and $k\,(x_k,\ y_k)$ in an anticlockwise sequence, which is a strictly rule we must obey in the following study. Each node has two coordinates $(x,\ y)$ which are known because they were fixed during meshing.

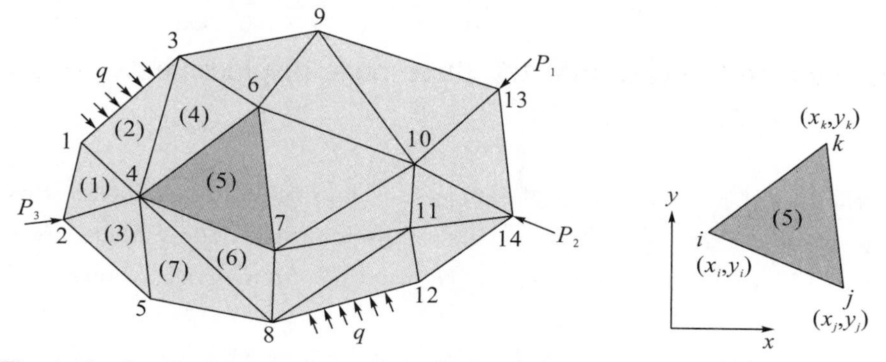

Fig. 1−3 One element is selected arbitrarily and its nodes are numbered by *i*, *j* and *k*

Let's think about: if one can find the solution of the stresses of the arbitrarily selected element, then one may take the same procedure to get the solutions of all the elements, which means you can obtain all the stresses and the strains of the whole domain or structure. Therefore, in the following study, we will first consider the simplest case only one element, which is also rational because one could use only one element to divide a domain in some simple situations. Then we will consider multi-elements, and through the relation between one element and multi-elements, one can easily get the solutions of stresses and strains of the structures.

Based on one element, two cases will be considered in this chapter: (1) the displacements of the three nodes are known, as shown in Fig. 1−4; (2) the loads acting on the three nodes are known, as shown in Fig. 1−5.

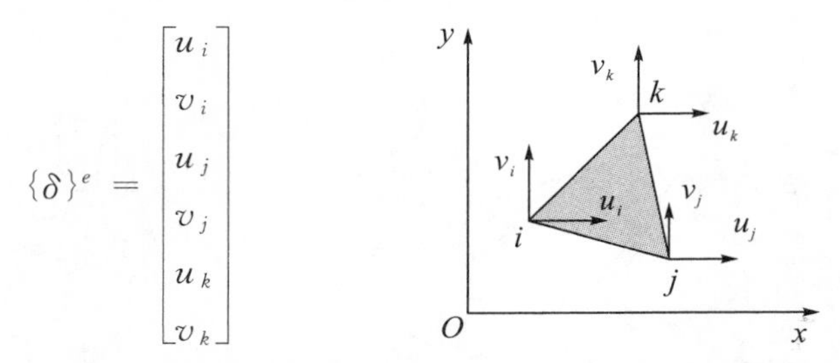

Fig. 1−4 The displacements of three nodes are known

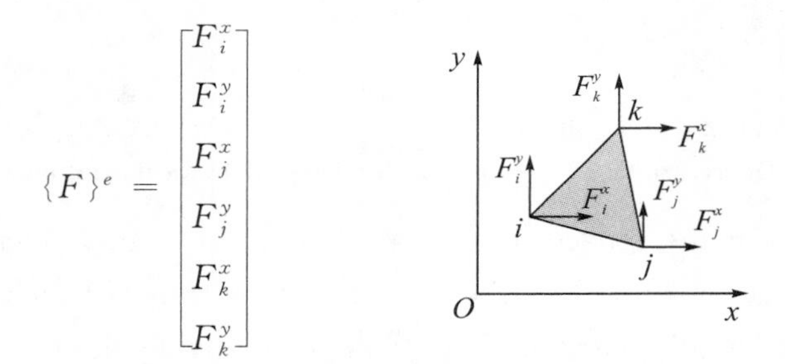

Fig. 1−5 The loads of three nodes are known

In the following, we will present detailed procedures of solving the stresses and strains for the two cases shown in Figs. 1−4 and 1−5.

1.2 Solution to the case that the three-node displacements are known

The case that the three-node displacements are known will be studied in this chapter. For each node, there are two displacements, thus for three nodes, there are totally six displacements $(\delta)^e$, and they can be expressed in terms of a matrix as shown in Fig. 1−6.

$$\{\delta\}^e = \begin{bmatrix} u_i \\ v_i \\ u_j \\ v_j \\ u_k \\ v_k \end{bmatrix}$$

Fig. 1-6 A triangle element and its six known displacement components

For the case that the displacements of the three nodes are known, we first try to obtain the displacements (u and v) of any point inside the triangular element based on the three node displacements. We use the strain-displacement relationship (i. e. $\varepsilon = \frac{\partial u}{\partial x}$) to obtain the strains of this element, and finally we use the stress-strain relationship (i. e. $\sigma_x = \frac{E}{1-\nu}(\varepsilon_x + \nu\varepsilon_y)$) to obtain the stresses of this element.

Next, we will follow this procedure to get the solution to the stresses of this element. We suppose the displacements insider this element are linearly related to the x and y coordinates, and the relation can be written as

$$\begin{aligned} u &= \alpha_1 + \alpha_2 x + \alpha_3 y \\ v &= \alpha_4 + \alpha_5 x + \alpha_6 y \end{aligned} \tag{1-1}$$

where α_1, α_2, α_3, α_4, α_5, α_6 are unknown coefficients. In a three-node element, there are totally six displacement components which are known and can be employed to determine the six unknown coefficients. Substituting the three node coordinates x and y into Eq. (1-1), one can get six linear simultaneous equations as

$$\begin{cases} u_i = \alpha_1 + \alpha_2 x_i + \alpha_3 y_i \\ u_j = \alpha_1 + \alpha_2 x_j + \alpha_3 y_j \\ u_k = \alpha_1 + \alpha_2 x_k + \alpha_3 y_k \end{cases} \Rightarrow \begin{Bmatrix} u_i \\ u_j \\ u_k \end{Bmatrix} = \begin{bmatrix} 1 & x_i & y_i \\ 1 & x_j & y_j \\ 1 & x_k & y_k \end{bmatrix} \begin{Bmatrix} \alpha_1 \\ \alpha_2 \\ \alpha_3 \end{Bmatrix} \tag{1-2}$$

$$\begin{cases} v_i = \alpha_4 + \alpha_5 x_i + \alpha_6 y_i \\ v_j = \alpha_4 + \alpha_5 x_j + \alpha_6 y_j \\ v_k = \alpha_4 + \alpha_5 x_k + \alpha_6 y_k \end{cases} \Rightarrow \begin{Bmatrix} v_i \\ v_j \\ v_k \end{Bmatrix} = \begin{bmatrix} 1 & x_i & y_i \\ 1 & x_j & y_j \\ 1 & x_k & y_k \end{bmatrix} \begin{Bmatrix} \alpha_4 \\ \alpha_5 \\ \alpha_6 \end{Bmatrix} \tag{1-3}$$

1.2.1 The area of a triangle element

In order to obtain the solution of the six unknown coefficients α_1, α_2, α_3, α_4, α_5, α_6, we will first investigate the determinant

$$D = \begin{vmatrix} 1 & x_i & y_i \\ 1 & x_j & y_j \\ 1 & x_k & y_k \end{vmatrix} = x_j y_k + x_k y_i + x_i y_j - x_j y_i - x_i y_k - x_k y_j \tag{1-4}$$

Combined with Fig. 1-7, Eq. (1-4) can be written in another form as

$$D = y_k(x_j - x_i) + y_j(x_i - x_k) + y_i(x_k - x_j) = 2A \tag{1-5}$$

where A is the triangle area in Fig. 1—7. This indicates that the determinant D is just the double triangle area.

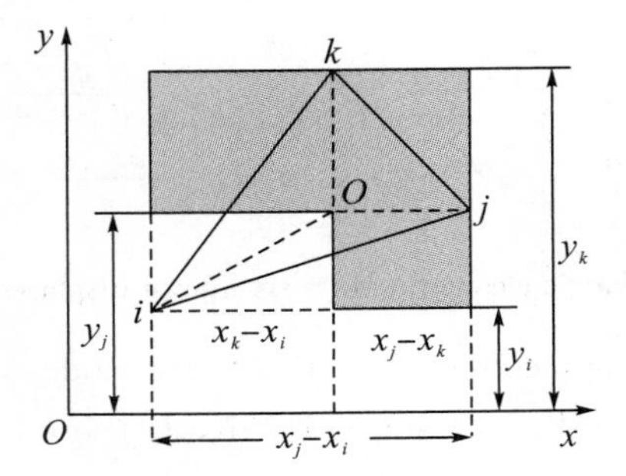

Fig. 1—7 Sketch of a triangle

1.2.2 The solution of the six unknown coefficients

According to the knowledge of Linear Algebra, from Eq. (1—2), the coefficients α_1, α_2 and α_3 can be expressed as

$$\alpha_1=\frac{1}{D}\begin{vmatrix} u_i & x_i & y_i \\ u_j & x_j & y_j \\ u_k & x_k & y_k \end{vmatrix},\ \alpha_2=\frac{1}{D}\begin{vmatrix} 1 & u_i & y_i \\ 1 & u_j & y_j \\ 1 & u_k & y_k \end{vmatrix},\ \alpha_3=\frac{1}{D}\begin{vmatrix} 1 & x_i & u_i \\ 1 & x_j & u_j \\ 1 & x_k & u_k \end{vmatrix} \tag{1—6}$$

where $D=\begin{vmatrix} 1 & x_i & y_i \\ 1 & x_j & y_j \\ 1 & x_k & y_k \end{vmatrix}=2A$, and more details of the coefficients α_1, α_2 and α_3 can be written as

$$\alpha_1=\frac{1}{D}\begin{vmatrix} u_i & x_i & y_i \\ u_j & x_j & y_j \\ u_k & x_k & y_k \end{vmatrix}=\frac{1}{2A}\left(u_i\begin{vmatrix} x_j & y_j \\ x_k & y_k \end{vmatrix}-u_j\begin{vmatrix} x_i & y_i \\ x_k & y_k \end{vmatrix}+u_k\begin{vmatrix} x_i & y_i \\ x_j & y_j \end{vmatrix}\right)$$

$$=\frac{1}{2A}(a_iu_i+a_ju_j+a_ku_k)$$

$$\alpha_2=\frac{1}{D}\begin{vmatrix} 1 & u_i & y_i \\ 1 & u_j & y_j \\ 1 & u_k & y_k \end{vmatrix}=\frac{1}{2A}\left(-u_i\begin{vmatrix} 1 & y_j \\ 1 & y_k \end{vmatrix}+u_j\begin{vmatrix} 1 & y_i \\ 1 & y_k \end{vmatrix}-u_k\begin{vmatrix} 1 & y_i \\ 1 & y_j \end{vmatrix}\right)$$

$$=\frac{1}{2A}(b_iu_i+b_ju_j+b_ku_k)$$

$$\alpha_3=\frac{1}{D}\begin{vmatrix} 1 & x_i & u_i \\ 1 & x_j & u_j \\ 1 & x_k & u_k \end{vmatrix}=\frac{1}{2A}\left(u_i\begin{vmatrix} 1 & x_j \\ 1 & x_k \end{vmatrix}-u_j\begin{vmatrix} 1 & x_i \\ 1 & x_k \end{vmatrix}+u_k\begin{vmatrix} 1 & x_i \\ 1 & x_j \end{vmatrix}\right)$$

$$=\frac{1}{2A}(c_iu_i+c_ju_j+c_ku_k)$$

where $a_i=\begin{vmatrix} x_j & y_j \\ x_k & y_k \end{vmatrix}=x_jy_k-x_ky_j$; $b_i=-\begin{vmatrix} 1 & y_j \\ 1 & y_k \end{vmatrix}=y_j-y_k$; $c_i=\begin{vmatrix} 1 & x_j \\ 1 & x_k \end{vmatrix}=-x_j+x_k$;

(i, j, k), and here the subscripts i, j and k circulate in the same manner. For example, according to the i, j and k sequence, as shown in Fig. 1−8, a_j can be written as $a_j = x_k y_i - x_i y_k$. Similarly, all the other coefficients a_k, b_j, b_k, c_j and c_k can be obtained, and the students should write out them as an assignment.

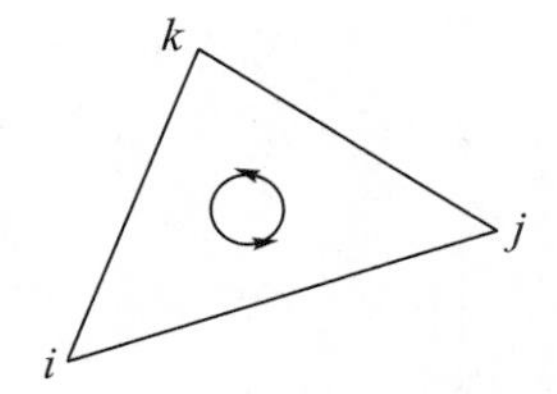

Fig. 1−8 Sketch of a triangle

It should be noted that all the a, b and c coefficients are constants because the nodal coordinates (x, y) are known and have a fixed value for a specific triangle element.

Similarly, from Eq. (1−3), one can gain the solution of the coefficients α_4, α_5 and α_6, and they can be expressed as

$$\alpha_4 = \frac{1}{2A}(a_i v_i + a_j v_j + a_k v_k)$$

$$\alpha_5 = \frac{1}{2A}(b_i v_i + b_j v_j + b_k v_k)$$

$$\alpha_6 = \frac{1}{2A}(c_i v_i + c_j v_j + c_k v_k)$$

It can be seen that the six coefficients α_1, α_2, α_3, α_4, α_5, α_6 are related to the node displacements (u_i, v_i) (i, j, k). The six coefficients can be written in another form as

$$\begin{Bmatrix} \alpha_1 \\ \alpha_2 \\ \alpha_3 \end{Bmatrix} = \frac{1}{2A}\begin{bmatrix} a_i & a_j & a_k \\ b_i & b_j & b_k \\ c_i & c_j & c_k \end{bmatrix}\begin{Bmatrix} u_i \\ u_j \\ u_k \end{Bmatrix}, \quad \begin{Bmatrix} \alpha_4 \\ \alpha_5 \\ \alpha_6 \end{Bmatrix} = \frac{1}{2A}\begin{bmatrix} a_i & a_j & a_k \\ b_i & b_j & b_k \\ c_i & c_j & c_k \end{bmatrix}\begin{Bmatrix} v_i \\ v_j \\ v_k \end{Bmatrix} \tag{1−7}$$

1.2.3 The displacement expression and shape function

Substituting the six coefficients in Eq. (1−7) into the displacement expression in Eq. (1−1), one can have

$$u = \alpha_1 + \alpha_2 x + \alpha_3 y = [1 \quad x \quad y]\begin{Bmatrix} \alpha_1 \\ \alpha_2 \\ \alpha_3 \end{Bmatrix} = \frac{1}{2A}[1 \quad x \quad y]\begin{bmatrix} a_i & a_j & a_k \\ b_i & b_j & b_k \\ c_i & c_j & c_k \end{bmatrix}\begin{Bmatrix} u_i \\ u_j \\ u_k \end{Bmatrix}$$

$$= \begin{bmatrix} N_i \\ N_j \\ N_k \end{bmatrix}^{\mathrm{T}}\begin{Bmatrix} u_i \\ u_j \\ u_k \end{Bmatrix} = N_i u_i + N_j u_j + N_k u_k$$

$$v = \alpha_4 + \alpha_5 x + \alpha_6 y = [1 \quad x \quad y]\begin{Bmatrix} \alpha_4 \\ \alpha_5 \\ \alpha_6 \end{Bmatrix} = \frac{1}{2A}[1 \quad x \quad y]\begin{bmatrix} a_i & a_j & a_k \\ b_i & b_j & b_k \\ c_i & c_j & c_k \end{bmatrix}\begin{Bmatrix} v_i \\ v_j \\ v_k \end{Bmatrix}$$

$$= \begin{bmatrix} N_i \\ N_j \\ N_k \end{bmatrix}^{\mathrm{T}} \begin{Bmatrix} v_i \\ v_j \\ v_k \end{Bmatrix} = N_i v_i + N_j v_j + N_k v_k \qquad (1-8)$$

where $\begin{Bmatrix} N_i \\ N_j \\ N_k \end{Bmatrix} = \frac{1}{2A}[1\ x\ y]\begin{bmatrix} a_i & a_j & a_k \\ b_i & b_j & b_k \\ c_i & c_j & c_k \end{bmatrix} = \frac{1}{2A}\begin{Bmatrix} a_i + b_i x + c_i y \\ a_j + b_j x + c_j y \\ a_k + b_k x + c_k y \end{Bmatrix}$; and where $\begin{Bmatrix} a_i = x_j y_k - x_k y_j \\ b_i = y_j - y_k \\ c_i = -x_j + x_k \end{Bmatrix}$

(i, j, k)

where N_i, N_j and N_k are called Shape Function of this triangle element, and for simplification, they can be written as

$$N_i = \frac{1}{2A}(a_i + b_i x + c_i y) \qquad (i, j, k) \qquad (1-9)$$

For a triangular element, the area A and the coefficients a_i, b_i and c_i are constants, and therefore, the shape function is only the function of the coordinates x and y.

The displacements in Eq. (1−8) can be simply expressed as

$$\begin{Bmatrix} u \\ v \end{Bmatrix} = \begin{Bmatrix} N_i u_i + N_j u_j + N_k u_k \\ N_i v_i + N_j v_j + N_k v_k \end{Bmatrix} = \begin{bmatrix} N_i & 0 & N_j & 0 & N_k & 0 \\ 0 & N_i & 0 & N_j & 0 & N_k \end{bmatrix} \begin{Bmatrix} u_i \\ v_i \\ u_j \\ v_j \\ u_k \\ v_k \end{Bmatrix} = [N]\{\delta\}^e$$

$$(1-10)$$

where $[N]$ is the matrix of shape function. It can be seen that the displacements within each element and on its boundaries can be expressed in terms of the shape function and its nodal displacements.

1.2.4 Strain and strain matrix

After the displacement functions of the element have been determined, we can use the general relation of strain-displacement to get the strains inside this element. The relation of strain-displacement for a plane problem can be expressed as

$$\{\varepsilon\} = \begin{Bmatrix} \varepsilon_x \\ \varepsilon_y \\ \gamma_{xy} \end{Bmatrix} = \begin{Bmatrix} \frac{\partial u}{\partial x} \\ \frac{\partial v}{\partial y} \\ \frac{\partial u}{\partial y} + \frac{\partial v}{\partial x} \end{Bmatrix} = \begin{bmatrix} \frac{\partial}{\partial x} & 0 \\ 0 & \frac{\partial}{\partial y} \\ \frac{\partial}{\partial y} & \frac{\partial}{\partial x} \end{bmatrix} \begin{Bmatrix} u \\ v \end{Bmatrix} \qquad (1-11)$$

Substituting Eq. (1−10) into Eq. (1−11), we have

$$\{\varepsilon\}=\begin{bmatrix}\frac{\partial}{\partial x} & 0\\ 0 & \frac{\partial}{\partial y}\\ \frac{\partial}{\partial y} & \frac{\partial}{\partial x}\end{bmatrix}\begin{bmatrix}N_i & 0 & N_j & 0 & N_k & 0\\ 0 & N_i & 0 & N_j & 0 & N_k\end{bmatrix}\begin{Bmatrix}u_i\\ v_i\\ u_j\\ v_j\\ u_k\\ v_k\end{Bmatrix}=\begin{bmatrix}\frac{\partial N_i}{\partial x} & 0 & \frac{\partial N_j}{\partial x} & 0 & \frac{\partial N_k}{\partial x} & 0\\ 0 & \frac{\partial N_i}{\partial y} & 0 & \frac{\partial N_j}{\partial y} & 0 & \frac{\partial N_k}{\partial y}\\ \frac{\partial N_i}{\partial y} & \frac{\partial N_i}{\partial x} & \frac{\partial N_j}{\partial y} & \frac{\partial N_j}{\partial x} & \frac{\partial N_k}{\partial y} & \frac{\partial N_k}{\partial x}\end{bmatrix}\begin{Bmatrix}u_i\\ v_i\\ u_j\\ v_j\\ u_k\\ v_k\end{Bmatrix}$$

Substituting the shape function $N_i=\frac{1}{2A}(a_i+b_ix+c_iy)(i,j,k)$ into the above equation, we have

$$\{\varepsilon\}=\begin{Bmatrix}\varepsilon_x\\ \varepsilon_y\\ \gamma_{xy}\end{Bmatrix}=\frac{1}{2A}\begin{bmatrix}b_i & 0 & b_j & 0 & b_k & 0\\ 0 & c_i & 0 & c_j & 0 & c_k\\ c_i & b_i & c_j & b_j & c_k & b_k\end{bmatrix}\begin{Bmatrix}u_i\\ v_i\\ u_j\\ v_j\\ u_k\\ v_k\end{Bmatrix}=[B]\{\delta\}^e \qquad (1-12)$$

where $b_i=y_j-y_k, c_i=-x_j+x_k(i,j,k)$, $[B]$ is strain matrix, and $(\delta)^e$ is element displacements. It can be seen that the matrix $[B]$ is a constant matrix, which means that the strains inside the whole element are the same because no x and y are related in the $[B]$ matrix.

One can find that if the displacements of the three nodes are known, i. e. $(\delta)^e$ is known, the three strain components can be easily obtained by using Eq. (1−12). The $[B]$ matrix can be written in another form as

$$[B]=[B_i \quad B_j \quad B_k]=\frac{1}{2A}\begin{bmatrix}b_i & 0 & b_j & 0 & b_k & 0\\ 0 & c_i & 0 & c_j & 0 & c_k\\ c_i & b_i & c_j & b_j & c_k & b_k\end{bmatrix} \qquad (1-13)$$

where $[B_i]=\frac{1}{2A}\begin{bmatrix}b_i & 0\\ 0 & c_i\\ c_i & b_i\end{bmatrix}(i,j,k)$.

1.2.5 Stresses and stress matrix

The relation between stresses and strains is the basic knowledge, and the students should have grasped this relationship before. It can be written as

$$\begin{cases}\sigma_x=\frac{E}{1-\nu^2}(\varepsilon_x+\nu\varepsilon_y)\\ \sigma_y=\frac{E}{1-\nu^2}(\nu\varepsilon_x+\varepsilon_y)\\ \tau_{xy}=\frac{E}{2(1+\nu)}\gamma_{xy}=\frac{E}{1-\nu^2}\cdot\frac{1-\nu}{2}\gamma_{xy}\end{cases}$$

The stress-strain relation can be rewritten in another form as

$$\begin{Bmatrix} \sigma_x \\ \sigma_y \\ \tau_{xy} \end{Bmatrix} = \frac{E}{1-\nu^2} \begin{bmatrix} 1 & \nu & 0 \\ \nu & 1 & 0 \\ 0 & 0 & \frac{1-\nu}{2} \end{bmatrix} \begin{Bmatrix} \varepsilon_x \\ \varepsilon_y \\ \gamma_{xy} \end{Bmatrix} \tag{1-14}$$

Eq. (1-14) can be simplified as

$$\{\sigma\} = [D]\{\varepsilon\} \tag{1-15}$$

where $[D]$ is the plane stress elasticity matrix and

$$[D] = \frac{E}{1-\nu^2} \begin{bmatrix} 1 & \nu & 0 \\ \nu & 1 & 0 \\ 0 & 0 & \frac{1-\nu}{2} \end{bmatrix} \tag{1-16}$$

Substituting the strains in Eq. (1-12) into Eq. (1-15), we have

$$\{\sigma\} = [D]\{\varepsilon\} = [D][B]\{\delta\}^e = [S]\{\delta\}^e \tag{1-17}$$

where $[S]$ is stress matrix and it can be written as

$$[S] = [D][B] = \frac{E}{2A(1-\nu^2)} \begin{bmatrix} 1 & \nu & 0 \\ \nu & 1 & 0 \\ 0 & 0 & \frac{1-\nu}{2} \end{bmatrix} \begin{bmatrix} b_i & 0 & b_j & 0 & b_k & 0 \\ 0 & c_i & 0 & c_j & 0 & c_k \\ c_i & b_i & c_j & b_j & c_k & b_k \end{bmatrix}$$

$$= \frac{E}{2A(1-\nu^2)} \begin{bmatrix} b_i & \nu c_i & b_j & \nu c_j & b_k & \nu c_k \\ \nu b_i & c_i & \nu b_j & c_j & \nu b_k & c_k \\ \frac{1-\nu}{2}c_i & \frac{1-\nu}{2}b_i & \frac{1-\nu}{2}c_j & \frac{1-\nu}{2}b_j & \frac{1-\nu}{2}c_k & \frac{1-\nu}{2}b_k \end{bmatrix} \tag{1-18}$$

where $b_i = y_j - y_k, c_i = -x_j + x_k (i, j, k)$.

Substituting Eq. (1-18) into Eq. (1-17), one can have

$$\begin{Bmatrix} \sigma_x \\ \sigma_y \\ \tau_{xy} \end{Bmatrix} = \frac{E}{2A(1-\nu^2)} \begin{bmatrix} b_i & \nu c_i & b_j & \nu c_j & b_k & \nu c_k \\ \nu b_i & c_i & \nu b_j & c_j & \nu b_k & c_k \\ \frac{1-\nu}{2}c_i & \frac{1-\nu}{2}b_i & \frac{1-\nu}{2}c_j & \frac{1-\nu}{2}b_j & \frac{1-\nu}{2}c_k & \frac{1-\nu}{2}b_k \end{bmatrix} \begin{bmatrix} u_i \\ v_i \\ u_j \\ v_j \\ u_k \\ v_k \end{bmatrix} \tag{1-19}$$

It can be seen that the stress matrix is also a constant matrix, which means that the stresses in this whole element are the same. Therefore, the triangle element is also called constant stress elements or constant strain elements. This is because the displacements selected in Eq. (1-1) are linearly related to x and y coordinates, and the derivatives with respect to x and y are independent of x and y. Therefore, both the $[B]$ matrix and the $[S]$ matrix are constants.

1.2.6 Summary

From the above study, we can find that for the case that the displacements $(\delta)^e$ of the three nodes are known, both strains and stresses are related to the nodal displacements, and the relationship can be written as $\{\varepsilon\}=[B]\{\delta\}^e$ and $\{\sigma\}=[S]\{\delta\}^e$. If the displacements of the three nodes $(\delta)^e$ are known, then we can use the $[B]$ matrix to calculate the strains of this element, i. e.

$$\{\varepsilon\}=\begin{Bmatrix}\varepsilon_x\\ \varepsilon_y\\ \gamma_{xy}\end{Bmatrix}=\frac{1}{2A}\begin{bmatrix}b_i & 0 & b_j & 0 & b_k & 0\\ 0 & c_i & 0 & c_j & 0 & c_k\\ c_i & b_i & c_j & b_j & c_k & b_k\end{bmatrix}\begin{Bmatrix}u_i\\ v_i\\ u_j\\ v_j\\ u_k\\ v_k\end{Bmatrix}$$

$$\{\delta\}^e=\begin{bmatrix}u_i\\ v_i\\ u_j\\ v_j\\ u_k\\ v_k\end{bmatrix}$$

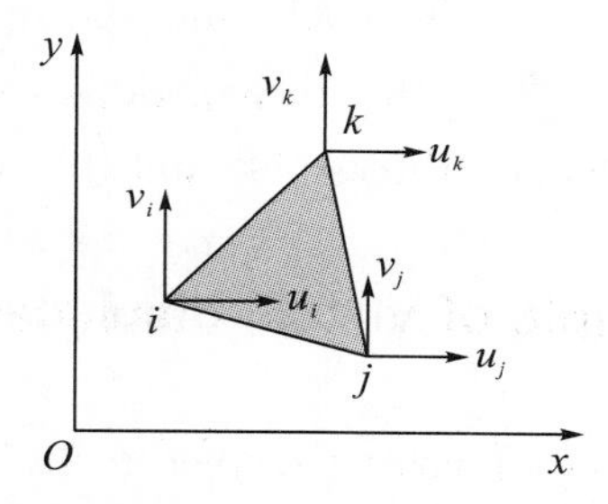

and we can use the $[S]$ matrix to calculate the stresses of this element, i. e.

$$\begin{Bmatrix}\sigma_x\\ \sigma_y\\ \tau_{xy}\end{Bmatrix}=\frac{E}{2A(1-\nu^2)}\begin{bmatrix}b_i & \nu c_i & b_j & \nu c_j & b_k & \nu c_k\\ \nu b_i & c_i & \nu b_j & c_j & \nu b_k & c_k\\ \frac{1-\nu}{2}c_i & \frac{1-\nu}{2}b_i & \frac{1-\nu}{2}c_j & \frac{1-\nu}{2}b_j & \frac{1-\nu}{2}c_k & \frac{1-\nu}{2}b_k\end{bmatrix}\begin{bmatrix}u_i\\ v_i\\ u_j\\ v_j\\ u_k\\ v_k\end{bmatrix}$$

It can be seen that for the case that the displacements of the three nodes are known, the strains can be readily calculated by the formula $\{\varepsilon\}=[B]\{\delta\}^e$, and the stresses can be easily calculated by $\{\sigma\}=[S]\{\delta\}^e$.

For the case that the loads of three nodes are known, which is more common phenomenon you may encounter in engineering practice, it will be studied in the follows.

1.3 Solutions to the case that the three-node loads are known

The case that three-node loads are known will be studied in this chapter. For each node, there are two loads in x and y direction, respectively, and for three nodes, there are totally six loads $\{F\}^e$, and they can be expressed as shown in Fig. 1—9.

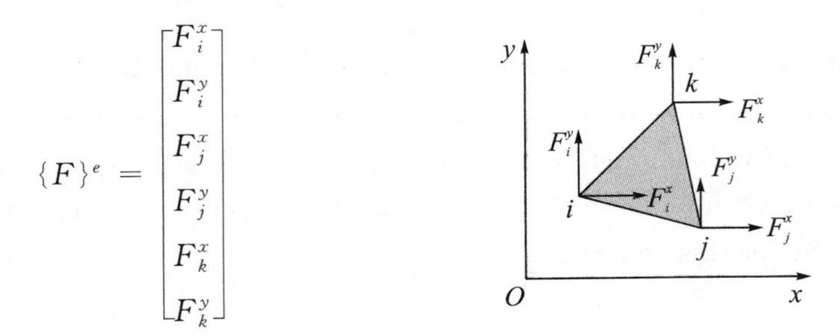

Fig. 1—9 A triangle element and the six loads

For the case that three-node displacements are known, we have learnt that the strains can be readily calculated by $\{\varepsilon\} = [B]\{\delta\}^e$, and the stresses can be easily calculated by $\{\sigma\} = [S]\{\delta\}^e$. For the case that three-node loads are known, we may try to find the relationship between the loads and the displacements; and from this relationship, the displacements $\{\delta\}^e$ of the three nodes can be obtained, and then we take the same procedure to calculate the strains and the stresses.

1.3.1 Principle of virtual displacement

The principle of virtual displacement state that: if a body in equilibrium under the applied loads is subjected to a kinematically admissible virtual displacement state, then the total internal virtual strain energy is equal to the total external virtual work.

For an element shown in Fig. 1—9, the virtual displacements and external loads are

$$\{\delta^*\}^e = \begin{bmatrix} u_i^* \\ v_i^* \\ u_j^* \\ v_j^* \\ u_k^* \\ v_k^* \end{bmatrix}, \qquad \{F\}^e = \begin{bmatrix} F_i^x \\ F_i^y \\ F_j^x \\ F_j^y \\ F_k^x \\ F_k^y \end{bmatrix} \tag{1—20}$$

Then the total external virtual work done by the external loads can be written as

$$(\{\delta^*\}^e)^{\mathrm{T}}\{F\}^e = F_i^x u_i^* + F_i^y v_i^* + F_j^x u_j^* + F_j^y v_j^* + F_k^x u_k^* + F_k^y v_k^* \tag{1—21}$$

For an element, the total strain energy induced by the virtual strains (due to incompatibility displacements) can be written as

$$\iint_A \{\varepsilon^*\}^{\mathrm{T}}\{\sigma\} t \, \mathrm{d}x \mathrm{d}y \tag{1—22}$$

where ε^* is virtual strain, and $\{\varepsilon^*\} = \begin{bmatrix} \varepsilon_x^* \\ \varepsilon_y^* \\ \gamma_{xy}^* \end{bmatrix}$; $\{\sigma\} = \begin{bmatrix} \sigma_x \\ \sigma_y \\ \tau_{xy} \end{bmatrix}$.

so

$$\{\varepsilon^*\}^{\mathrm{T}}\{\sigma\} = \sigma_x\varepsilon_x^* + \sigma_y\varepsilon_y^* + \tau_{xy}\gamma_{xy}^* \tag{1-23}$$

Discussion: According to material mechanics, usually the strain energy can be expressed as

$$\frac{1}{2}\sigma \cdot \varepsilon \tag{1-24}$$

The question is why there is no "1/2" in Eq. (1−22). In the course of material mechanics, you may study that for the bar under tensile loading, as shown in Fig. 1−10. The strain energy can be written as Eq. (1−24) because the stress-strain relation is the inclined straight line in Fig. 1−10. However, for the body in equilibrium, the stress already exists and is a constant during the virtual deformation, and therefore, there is no "1/2" in Eq. (1−22).

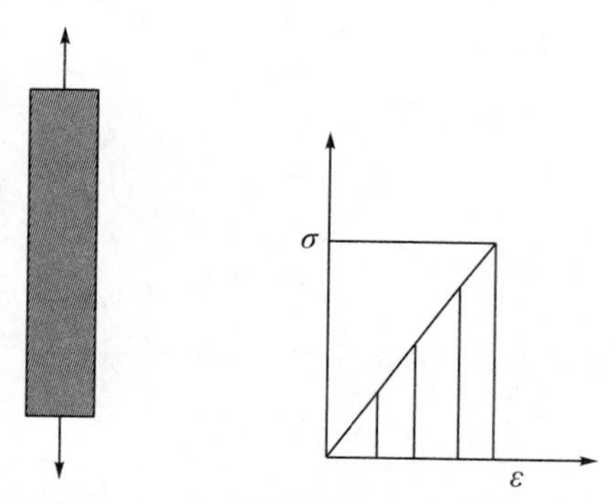

Fig. 1−10 A bar under tension and the stress-strain relationship

According to the virtual displacement principle, we have

$$(\{\delta^*\}^e)^{\mathrm{T}}\{F\}^e = \iint_A \{\varepsilon^*\}^{\mathrm{T}}\{\sigma\} t\,\mathrm{d}x\mathrm{d}y \tag{1-25}$$

where

$$\{\varepsilon^*\} = [B]\{\delta^*\} \text{ (due to } \{\varepsilon\} = [B]\{\delta\}) \tag{1-26}$$

$$\{\sigma\} = [D]\{\varepsilon\} = [D][B]\{\delta\}^e = [S]\{\delta\}^e \tag{1-27}$$

Substituting the strain and stress matrix into the above equation, we can have

$$(\{\delta^*\}^e)^{\mathrm{T}}\{F\}^e = \iint_A (\{\delta^*\}^e)^{\mathrm{T}}[B]^{\mathrm{T}}[D][B]\{\delta\}^e t\,\mathrm{d}x\mathrm{d}y \tag{1-28}$$

where $\{\delta^*\}^e$, $\{\delta\}^e$ are the three node displacements, and they are independent of x and y coordinate, and therefore, they can be moved out the integral, and then Eq. (1−28) can be rewritten as

$$(\{\delta^*\}^e)^{\mathrm{T}}\{F\}^e = (\{\delta^*\}^e)^{\mathrm{T}}\left(\iint_A [B]^{\mathrm{T}}[D][B] t\,\mathrm{d}x\mathrm{d}y\right)\{\delta\}^e \tag{1-29}$$

We delete $(\{\delta^*\}^e)^{\mathrm{T}}$, and then

$$\{F\}^e = \left(\iint_A [B]^{\mathrm{T}}[D][B] t\,\mathrm{d}x\mathrm{d}y\right)\{\delta\}^e \tag{1-30}$$

Eq. (1−30) can be simplified as

$$\{F\}^e = [K]^e\{\delta\}^e \tag{1-31}$$

where

$$[K]^e = \iint_A [B]^{\mathrm{T}}[D][B] t \,\mathrm{d}x\,\mathrm{d}y \tag{1-32}$$

where $[K]^e$ is the element stiffness matrix which will be analyzed in the following.

1.3.2 Element stiffness matrix

In the element stiffness matrixin Eq. (1−32), the integral is a constant because the matrix $[B]$ and $[D]$ are constants, and therefore, $[K]^e$ can be written as

$$[K]^e = [B]^{\mathrm{T}}[D][B]\cdot tA = [B]^{\mathrm{T}}[S]\cdot tA \tag{1-33}$$

where

$$[B] = \frac{1}{2A}\begin{bmatrix} b_i & 0 & b_j & 0 & b_k & 0 \\ 0 & c_i & 0 & c_j & 0 & c_k \\ c_i & b_i & c_j & b_j & c_k & b_k \end{bmatrix} \qquad \begin{cases} b_i = y_j - y_k \\ c_i = -x_j + x_k \end{cases} \quad (i, j, k)$$

[shown in Eq. (1−13)]

then

$$[B]^{\mathrm{T}} = \frac{1}{2A}\begin{bmatrix} b_i & 0 & c_i \\ 0 & c_i & b_i \\ b_j & 0 & c_j \\ 0 & c_j & b_j \\ b_k & 0 & c_k \\ 0 & c_k & b_k \end{bmatrix}$$

and

$$[S] = \frac{E}{2A(1-\nu^2)}\begin{bmatrix} b_i & \nu c_i & b_j & \nu c_j & b_k & \nu c_k \\ \nu b_i & c_i & \nu b_j & c_j & \nu b_k & c_k \\ \frac{1-\nu}{2}c_i & \frac{1-\nu}{2}b_i & \frac{1-\nu}{2}c_j & \frac{1-\nu}{2}b_j & \frac{1-\nu}{2}c_k & \frac{1-\nu}{2}b_k \end{bmatrix}$$

[shown in Eq. (1−18)]

Substituting $[B]^{\mathrm{T}}$ and $[S]$ into Eq. (1−33), we have

$$[K]^e = \frac{Et}{4(1-\nu^2)A}\begin{bmatrix}
b_ib_i + \frac{1-\nu}{2}c_ic_i & \nu b_ic_i + \frac{1-\nu}{2}c_ib_i & b_ib_j + \frac{1-\nu}{2}c_ic_j & \nu b_ic_j + \frac{1-\nu}{2}c_ib_j & b_ib_k + \frac{1-\nu}{2}c_ic_k & \nu b_ic_k + \frac{1-\nu}{2}c_ib_k \\
 & K_{ii} & & K_{ij} & & K_{ik} \\
\nu c_ib_i + \frac{1-\nu}{2}b_ic_i & c_ic_i + \frac{1-\nu}{2}b_ib_i & \nu c_ib_j + \frac{1-\nu}{2}b_ic_j & c_ic_j + \frac{1-\nu}{2}b_ib_j & \nu c_ib_k + \frac{1-\nu}{2}b_ic_k & c_ic_k + \frac{1-\nu}{2}b_ib_k \\
b_ib_j + \frac{1-\nu}{2}c_ic_j & \nu b_jc_i + \frac{1-\nu}{2}c_jb_i & b_jb_j + \frac{1-\nu}{2}c_jc_j & \nu b_jc_j + \frac{1-\nu}{2}c_jb_j & b_jb_k + \frac{1-\nu}{2}c_jc_k & \nu b_jc_k + \frac{1-\nu}{2}c_jb_k \\
 & K_{ji} & & K_{jj} & & K_{jk} \\
\nu c_jb_i + \frac{1-\nu}{2}b_jc_i & c_jc_i + \frac{1-\nu}{2}b_jb_i & \nu c_jb_j + \frac{1-\nu}{2}b_jc_j & c_jc_j + \frac{1-\nu}{2}b_jb_j & \nu c_jb_k + \frac{1-\nu}{2}b_jc_k & c_jc_k + \frac{1-\nu}{2}b_jb_k \\
b_kb_i + \frac{1-\nu}{2}c_kc_i & \nu b_kc_i + \frac{1-\nu}{2}c_kb_i & b_kb_j + \frac{1-\nu}{2}c_kc_j & \nu b_kc_j + \frac{1-\nu}{2}c_kb_j & b_kb_k + \frac{1-\nu}{2}c_kc_k & \nu b_kc_k + \frac{1-\nu}{2}c_kb_k \\
 & K_{ki} & & K_{kj} & & K_{kk} \\
\nu c_kb_i + \frac{1-\nu}{2}b_kc_i & c_kc_i + \frac{1-\nu}{2}b_kb_i & \nu c_kb_j + \frac{1-\nu}{2}b_kc_j & c_kc_j + \frac{1-\nu}{2}b_kb_j & \nu c_kb_k + \frac{1-\nu}{2}b_kc_k & c_kc_k + \frac{1-\nu}{2}b_kb
\end{bmatrix} \tag{1-34}$$

where $\begin{cases} b_i = y_j - y_k \\ c_i = -x_j + x_k \end{cases}$ (i, j, k). $[K]^e$ can be written in a simplified form as

$$[K]^e = \begin{bmatrix} K_{ii} & K_{ij} & K_{ik} \\ K_{ji} & K_{jj} & K_{jk} \\ K_{ki} & K_{kj} & K_{kk} \end{bmatrix} \tag{1-35}$$

where

$$[K_{rs}] = \frac{Et}{4(1-\nu^2)A}\begin{bmatrix} b_r b_s + \frac{1-\nu}{2}c_r c_s & \nu b_r c_s + \frac{1-\nu}{2}c_r b_s \\ \nu c_r b_s + \frac{1-\nu}{2}b_r c_s & c_r c_s + \frac{1-\nu}{2}b_r b_s \end{bmatrix} \quad (r = i, j, k; s = i, j, k)$$

Therefore, the element equilibrium equation can be written in three forms

$$\begin{bmatrix} F_i^x \\ F_i^y \\ F_j^x \\ F_j^y \\ F_k^x \\ F_k^y \end{bmatrix} = \frac{Et}{4(1-\nu^2)A}\begin{bmatrix} b_i b_i + \frac{1-\nu}{2}c_i c_i & \nu b_i c_i + \frac{1-\nu}{2}c_i b_i & b_i b_j + \frac{1-\nu}{2}c_i c_j & \nu b_i c_j + \frac{1-\nu}{2}c_i b_j & b_i b_k + \frac{1-\nu}{2}c_i c_k & \nu b_i c_k + \frac{1-\nu}{2}c_i b_k \\ & K_{ii} & & K_{ij} & & K_{ik} \\ \nu c_i b_i + \frac{1-\nu}{2}b_i c_i & c_i c_i + \frac{1-\nu}{2}b_i b_i & \nu c_i b_j + \frac{1-\nu}{2}b_i c_j & c_i c_j + \frac{1-\nu}{2}b_i b_j & \nu c_i b_k + \frac{1-\nu}{2}b_i c_k & c_i c_k + \frac{1-\nu}{2}b_i b_k \\ b_i b_j + \frac{1-\nu}{2}c_i c_j & \nu b_j c_i + \frac{1-\nu}{2}c_j b_i & b_j b_j + \frac{1-\nu}{2}c_j c_j & \nu b_j c_j + \frac{1-\nu}{2}c_j b_j & b_j b_k + \frac{1-\nu}{2}c_j c_k & \nu b_j c_k + \frac{1-\nu}{2}c_j b_k \\ & K_{ji} & & K_{jj} & & K_{jk} \\ \nu c_j b_i + \frac{1-\nu}{2}b_j c_i & c_j c_i + \frac{1-\nu}{2}b_j b_i & \nu c_j b_j + \frac{1-\nu}{2}b_j c_j & c_j c_j + \frac{1-\nu}{2}b_j b_j & \nu c_j b_k + \frac{1-\nu}{2}b_j c_k & c_j c_k + \frac{1-\nu}{2}b_j b_k \\ b_k b_i + \frac{1-\nu}{2}c_k c_i & \nu b_k c_i + \frac{1-\nu}{2}c_k b_i & b_k b_j + \frac{1-\nu}{2}c_k c_j & \nu b_k c_j + \frac{1-\nu}{2}c_k b_j & b_k b_k + \frac{1-\nu}{2}c_k c_k & \nu b_k c_k + \frac{1-\nu}{2}c_k b_k \\ & K_{ki} & & K_{kj} & & K_{kk} \\ \nu c_k b_i + \frac{1-\nu}{2}b_k c_i & c_k c_i + \frac{1-\nu}{2}b_k b_i & \nu c_k b_j + \frac{1-\nu}{2}b_k c_j & c_k c_j + \frac{1-\nu}{2}b_k b_j & \nu c_k b_k + \frac{1-\nu}{2}b_k c_k & c_k c_k + \frac{1-\nu}{2}b_k b_k \end{bmatrix}\begin{bmatrix} u_i \\ v_i \\ u_j \\ v_j \\ u_k \\ v_k \end{bmatrix} \tag{1-36}$$

$$\begin{Bmatrix} \{F_i\} \\ \{F_j\} \\ \{F_k\} \end{Bmatrix} = \begin{bmatrix} K_{ii} & K_{ij} & K_{ik} \\ K_{ji} & K_{jj} & K_{jk} \\ K_{ki} & K_{kj} & K_{kk} \end{bmatrix}\begin{Bmatrix} \{\delta_i\} \\ \{\delta_j\} \\ \{\delta_k\} \end{Bmatrix} \tag{1-37}$$

and

$$\{F\}^e = [K]^e \{\delta\}^e \qquad \text{[shown in Eq. (1-31)]}$$

Because the loads are known, the six displacement components can be obtained by solving the six linear simultaneous equations in Eq. (1-36).

If a triangular zone is just discretized into only one element, and in the case that the three node loads are known, we can easily obtain the six displacement components, and then we use $[B]$ matrix to calculate the three strain components

$$\{\varepsilon\} = \begin{Bmatrix} \varepsilon_x \\ \varepsilon_y \\ \gamma_{xy} \end{Bmatrix} = \frac{1}{2A}\begin{bmatrix} b_i & 0 & b_j & 0 & b_k & 0 \\ 0 & c_i & 0 & c_j & 0 & c_k \\ c_i & b_i & c_j & b_j & c_k & b_k \end{bmatrix}\begin{Bmatrix} u_i \\ v_i \\ u_j \\ v_j \\ u_k \\ v_k \end{Bmatrix} \qquad \text{[shown in Eq. (1-12)]}$$

and we can use the stress matrix $[S]$ to calculate the three stress components

$$\begin{Bmatrix}\sigma_x\\ \sigma_y\\ \tau_{xy}\end{Bmatrix}=\frac{E}{2A(1-\nu^2)}\begin{bmatrix} b_i & \nu c_i & b_j & \nu c_j & b_k & \nu c_k\\ \nu b_i & c_i & \nu b_j & c_j & \nu b_k & c_k\\ \frac{1-\nu}{2}c_i & \frac{1-\nu}{2}b_i & \frac{1-\nu}{2}c_j & \frac{1-\nu}{2}b_j & \frac{1-\nu}{2}c_k & \frac{1-\nu}{2}b_k\end{bmatrix}\begin{bmatrix}u_i\\ v_i\\ u_j\\ v_j\\ u_k\\ v_k\end{bmatrix}$$

[shown in Eq. (1−19)]

1.3.3 The properties of element stiffness matrix

The properties of element stiffness matrix can be summarized as follows:

(1) The element stiffness matrix describes the anti-deformation capability of an element, as the stiffness of a spring;

(2) The element stiffness matrix doesn't have relation with the element position (rigid movement);

(3) The value of K_{ii}, K_{jj} and K_{kk} in the diagonal must be positive in the element matrix;

(4) The element stiffness matrix is symmetric about the diagonal.

$$[K]^e=\frac{Et}{4(1-\nu^2)A}\left[\begin{array}{cccccc}
b_ib_i+\frac{1-\nu}{2}c_ic_i & \nu b_ic_i+\frac{1-\nu}{2}c_ib_i & b_ib_j+\frac{1-\nu}{2}c_ic_j & \nu b_ic_j+\frac{1-\nu}{2}c_ib_j & b_ib_k+\frac{1-\nu}{2}c_ic_k & \nu b_ic_k+\frac{1-\nu}{2}c_ib_k\\
 & K_{ii} & & K_{ij} & & K_{ik}\\
\nu c_ib_i+\frac{1-\nu}{2}b_ic_i & c_ic_i+\frac{1-\nu}{2}b_ib_i & \nu c_ib_j+\frac{1-\nu}{2}b_ic_j & c_ic_j+\frac{1-\nu}{2}b_ib_j & \nu c_ib_k+\frac{1-\nu}{2}b_ic_k & c_ic_k+\frac{1-\nu}{2}b_ib_k\\
b_ib_j+\frac{1-\nu}{2}c_ic_j & \nu b_jc_i+\frac{1-\nu}{2}c_jb_i & b_jb_j+\frac{1-\nu}{2}c_jc_j & \nu b_jc_j+\frac{1-\nu}{2}c_jb_j & b_jb_k+\frac{1-\nu}{2}c_jc_k & \nu b_jc_k+\frac{1-\nu}{2}c_jb_k\\
 & K_{ji} & & K_{jj} & & K_{jk}\\
\nu c_jb_i+\frac{1-\nu}{2}b_jc_i & c_jc_i+\frac{1-\nu}{2}b_jb_i & \nu c_jb_j+\frac{1-\nu}{2}b_jc_j & c_jc_j+\frac{1-\nu}{2}b_jb_j & \nu c_jb_k+\frac{1-\nu}{2}b_jc_k & c_jc_k+\frac{1-\nu}{2}b_jb_k\\
b_kb_i+\frac{1-\nu}{2}c_kc_i & \nu b_kc_i+\frac{1-\nu}{2}c_kb_i & b_kb_j+\frac{1-\nu}{2}c_kc_j & \nu b_kc_j+\frac{1-\nu}{2}c_kb_j & b_kb_k+\frac{1-\nu}{2}c_kc_k & \nu b_kc_k+\frac{1-\nu}{2}c_kb_k\\
 & K_{ki} & & K_{kj} & & K_{kk}\\
\nu c_kb_i+\frac{1-\nu}{2}b_kc_i & c_kc_i+\frac{1-\nu}{2}b_kb_i & \nu c_kb_j+\frac{1-\nu}{2}b_kc_j & c_kc_j+\frac{1-\nu}{2}b_kb_j & \nu c_kb_k+\frac{1-\nu}{2}b_kc_k & c_kc_k+\frac{1-\nu}{2}b_kb_k
\end{array}\right]$$

A calculate example

A triangular element as shown in Fig. 1−11, the plate thickness $t=1$; $E=2\times10^5$; $v=0.2$; calculate the element stiffness matrix.

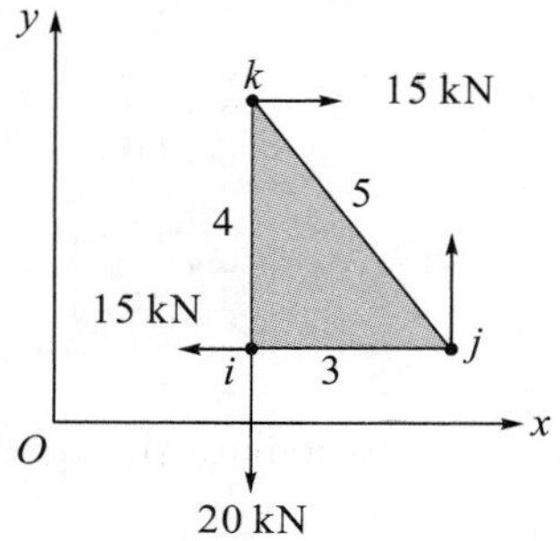

Fig. 1−11 A triangular element under loads

According to $\begin{cases} b_i = y_j - y_k \\ c_i = -x_j + x_k \end{cases}$ (i, j, k), we have

$$b_i = -4;\ b_j = y_k - y_i = 4;\ b_k = y_i - y_j = 0$$

$$c_i = -3;\ c_j = x_i - x_k = 0;\ c_k = x_j - x_i = 3$$

Submitting all the parameters into Eq. (1−34), one can have that

$$[K]^e = \begin{bmatrix} 19.6 & 7.2 & -16 & -4.8 & -3.6 & -2.4 \\ 7.2 & 15.4 & -2.4 & -6.4 & -4.8 & -9 \\ -16 & -2.4 & 16 & 0 & 0 & 2.4 \\ -4.8 & -6.4 & 0 & 6.4 & 4.8 & 0 \\ -3.6 & -4.8 & 0 & 4.8 & 3.6 & 0 \\ -2.4 & -9 & 2.4 & 0 & 0 & 9 \end{bmatrix}$$

The equilibrium equation, i. e. Eq. (1−36), can be written as

$$\begin{bmatrix} 19.6 & 7.2 & -16 & -4.8 & -3.6 & -2.4 \\ 7.2 & 15.4 & -2.4 & -6.4 & -4.8 & -9 \\ -16 & -2.4 & 16 & 0 & 0 & 2.4 \\ -4.8 & -6.4 & 0 & 6.4 & 4.8 & 0 \\ -3.6 & -4.8 & 0 & 4.8 & 3.6 & 0 \\ -2.4 & -9 & 2.4 & 0 & 0 & 9 \end{bmatrix} \begin{Bmatrix} u_i \\ v_i \\ u_j \\ v_j \\ u_k \\ v_k \end{Bmatrix} = \begin{Bmatrix} F_i^x \\ F_i^y \\ F_j^x \\ F_j^y \\ F_k^x \\ F_k^y \end{Bmatrix} = \begin{Bmatrix} -15 \\ -20 \\ 0 \\ 20 \\ 15 \\ 0 \end{Bmatrix}$$

From this equation, we can obtain the six displacements, and then use the $[B]$ matrix to calculate the strains, and use the $[S]$ matrix to calculate the stresses. The students are required to complete the subsequent calculation.

1.3.4 The properties of shape function

(1) For any point inside an element, $N_i + N_j + N_k = 1$.

According to the definition of shape function, we have

$$\begin{Bmatrix} N_i \\ N_j \\ N_k \end{Bmatrix} = \frac{1}{2A} \begin{Bmatrix} a_i + b_i x + c_i y \\ a_j + b_j x + c_j y \\ a_k + b_k x + c_k y \end{Bmatrix} \quad \text{[shown in Eq. (1−9)]}$$

And then

$$N_i + N_j + N_k = \frac{1}{2A}[a_i + a_j + a_k + (b_i + b_j + b_k)x + (c_i + c_j + c_k)y]$$

Because $a_i + a_j + a_k = x_j y_k - x_k y_j + x_k y_i - x_i y_k + x_i y_j - x_j y_i$, and

$$D = \begin{vmatrix} 1 & x_i & y_i \\ 1 & x_j & y_j \\ 1 & x_k & y_k \end{vmatrix} = x_j y_k + x_k y_i + x_i y_j - x_j y_i - x_i y_k - x_k y_j = 2A$$

[shown in Eq. (1−4)]

We have $a_i + a_j + a_k = 2A$. Because

$$b_i + b_j + b_k = 0, \quad c_i + c_j + c_k = 0$$

So we have $N_i + N_j + N_k = 1$.

(2) At the node i, $N_i = 1, N_j = N_k = 0$,

At the node j, $N_i = 0$, $N_j = 1, N_k = 0$,

At the node k, $N_i = 0$, $N_j = 0, N_k = 1$.

Because at node i ($x = x_i, y = y_i$)

$$N_i = \frac{1}{2A}[(x_j y_k - x_k y_j) + (y_j - y_k)x_i + (-x_j + x_k)y_i]$$

$$= \frac{1}{2A}(x_j y_k - x_k y_j + x_k y_i - x_i y_k + x_i y_j - x_j y_i)$$

$$= \frac{2A}{2A} = 1$$

Similar conclusions can be obtained for those at nodes j and k

$$N_j = \frac{1}{2A}[(x_k y_i - x_i y_k) + (y_k - y_i)x_i + (-x_k + x_i)y_i] = \frac{1}{2A}[0] = 0$$

$$N_k = \frac{1}{2A}[(x_i y_j - x_j y_i) + (y_i - y_j)x_i + (-x_i + x_j)y_i] = \frac{1}{2A}[0] = 0$$

1.3.5 The properties of displacement function

(1) Displacement function must contain a constant strain because when the element size is very small, the stain inside the element approaches a constant value which will become a dominant factor. For triangular elements, the displacement function is

$$\begin{aligned} u &= \alpha_1 + \alpha_2 x + \alpha_3 y \\ v &= \alpha_4 + \alpha_5 x + \alpha_6 y \end{aligned} \qquad \text{[shown in Eq. (1-1)]}$$

The strains are

$$\begin{cases} \varepsilon_x = \dfrac{\partial u}{\partial x} = \dfrac{\partial}{\partial x}(\alpha_1 + \alpha_2 x + \alpha_3 y) = \alpha_2 \\ \varepsilon_y = \dfrac{\partial v}{\partial y} = \dfrac{\partial}{\partial y}(\alpha_4 + \alpha_5 x + \alpha_6 y) = \alpha_6 \\ \gamma_{xy} = \dfrac{\partial u}{\partial y} + \dfrac{\partial v}{\partial x} = \alpha_3 + \alpha_5 \end{cases}$$

Because $\alpha_2, \alpha_3, \alpha_5, \alpha_6$ are constants, so the three strain components also constants.

(2) Displacement function must contain rigid displacements because the displacements usually contain rigid movement. The strains for triangle elements can be written as

$$\begin{cases} \varepsilon_x = \alpha_2 \\ \varepsilon_y = \alpha_6 \\ \gamma_{xy} = \alpha_3 + \alpha_5 \end{cases}$$

As the strain components $\varepsilon_x, \varepsilon_y, \gamma_{xy}$ equal zero, i. e. $\alpha_2 = \alpha_6 = \alpha_5 + \alpha_3 = 0$, then the displacement is

$$u = \alpha_1 + \alpha_2 x + \frac{\alpha_3 + \alpha_5}{2} y + \frac{\alpha_3 - \alpha_5}{2} y$$

$$v = \alpha_4 + \alpha_6 y + \frac{\alpha_3 + \alpha_5}{2} x + \frac{\alpha_5 - \alpha_3}{2} x$$

$$\Rightarrow \begin{cases} u = \alpha_1 + \dfrac{\alpha_3 - \alpha_5}{2} y \\ v = \alpha_4 + \dfrac{\alpha_5 - \alpha_3}{2} x \end{cases}$$

It can be seen that displacement function contains rigid displacements.

(3) Displacement compatibility is the key issue for choosing displacement function in the finite element method. If the displacement function cannot satisfy the compatibility requirements, the calculation results is not reliable.

For triangular elements, the displacement function is linear, and the displacements on the boundary AB are distributed on the line $A'B'$ connecting the two nodal displacements, as shown in Fig. 1—12. Therefore, the displacements occurring in the mutual boundary AB of the two adjacent elements are equal, which can make the boundary continuous and compatible, avoiding the following two cases of dislocations as shown in Fig. 1—12 occurring.

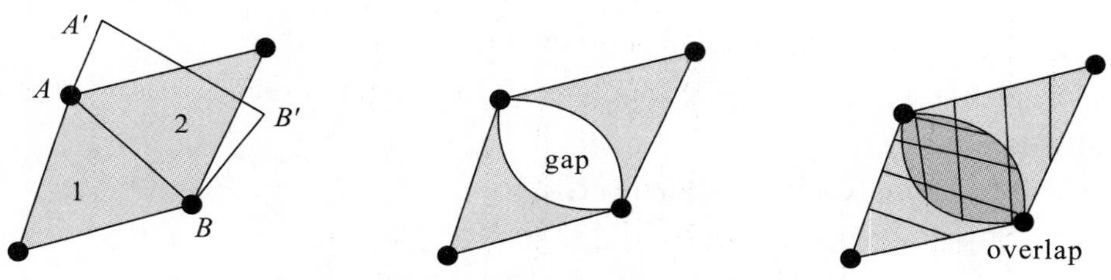

Fig. 1—12 Two adjacent elements and two cases of dislocations

Assignments:

1. The finite element method is actually a displacement method,

(1) Which basic equations of elastic mechanics are applied?

(2) How the deformation compatibility conditions are satisfied?

2. There is an isosceles right triangle element as shown in the following figure, where the Poisson's ratio is 0.25, Young modulus is 200 GPa, thickness of the plate is 1. Show the shape function matrix $[N]$, strain matrix $[B]$, stress matrix $[S]$ and element stiffness matrix $[K]^e$.

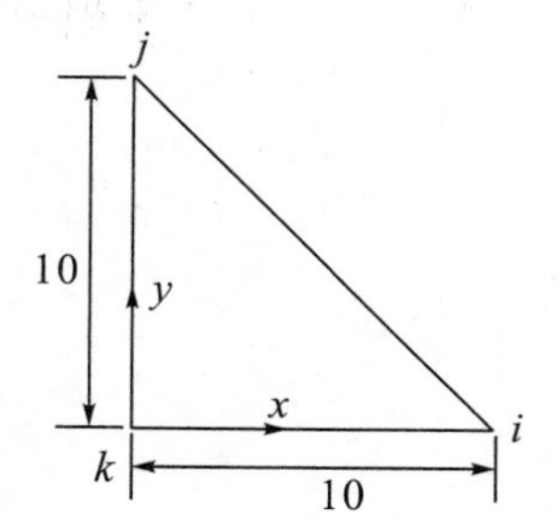

Chapter 2 Subroutine to calculate element stiffness matrix

It is necessary for students to design independently a finite element code by using either MATLAB or C programming language. The students without designing a finite element code cannot fully grasp the principle of the finite element method because the algebra and the matrix involved in the finite element method may appear somewhat tedious. On the other hand, with the fast development of modern high technology, the students are required to be able to write program codes either.

First we define some important variables and two arrays that are used to store two important parameter:

MNE is Maximum Number of Element;

MNN is Maximum Number of Nodes;

E0 is elastic modulus;

U0 is Poisson's ratio;

NNE(MNE, 3) is an array to store the three Node Numbers of each Elements, where MNE is the maximum number of element. In Fig. 2－1, the maximum number of elements MNE = 17, then maximum dimension should be NNE(17, 3). For the 5th element, $i=4$, $j=7$ and $k=6$, thus NNE(5, 1)=4, and NNE(5, 3)=6.

CN (MNN, 2) is an array to store the Coordinates of Nodes, where MNN is maximum number of nodes. In Fig. 2－1, the maximum number of nodes MNN=14, thus the maximum dimension is CN(14, 2). For example, CN(4, 1) denotes the x coordinate of node 4, and CN(4, 2) denotes the y coordinate of node 4.

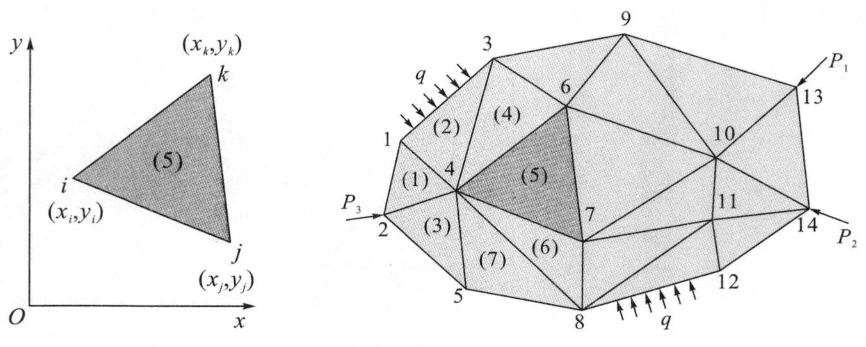

Fig. 2－1 Triangle elements

We use a subroutine (a function in C programming language or MATLAB) to calculate element stiffness matrix, and the subroutine is named ESM. Some parameters (information) are transmitted through the variables in the bracket following the subroutine name. The dimension of arrays is defined in the second row, and all the arrays should be defined here.

```
Subroutine ESM(NE, NNE, CN, E0, U0, t, MNE, MNN, EK, S)
Dimension NNE(MNE, 3), CN(MNN, 2), B(3, 6), D(3, 3), S(3, 6), EK(6, 6)
```

where t is plate thickness, B, S, D and EK are strain matrix, stress matrix, elasticity matrix and element stiffness matrix, respectively.

2.1 Calculating triangle area

Recall the element area calculation formula

$$D = \begin{vmatrix} 1 & x_i & y_i \\ 1 & x_j & y_j \\ 1 & x_k & y_k \end{vmatrix} = x_j y_k + x_k y_i + x_i y_j - x_j y_i - x_i y_k - x_k y_j$$

$$= y_k(x_j - x_i) + y_j(x_i - x_k) + y_i(x_k - x_j) = 2A$$

[shown in Eq. (1-5)]

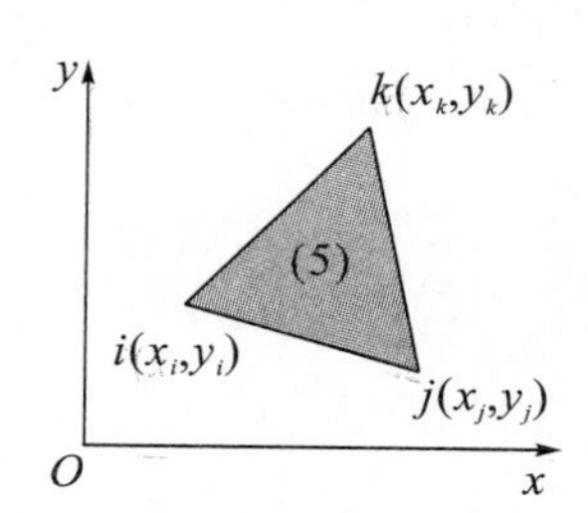

Fig. 2-2 A triangle element

According the above equation and the triangle shown in Fig. 2-2, we can write the program code as

```
i=NNE(NE, 1)
j=NNE(NE, 2)
k=NNE(NE, 3)
Xji=CN(j, 1)-CN(i, 1)        ➡    [note: Xji=xj-xi]
Xik=CN(i, 1)-CN(k, 1)        ➡    [note: Xik=xi-xk]
Xkj=CN(k, 1)-CN(j, 1)        ➡    [note: Xkj=xk-xj]
Area=0.5 * [CN(k, 2) * Xji+CN(j, 2) * Xik+CN(i, 2)] * Xkj
```

2.2 Calculating [B] strain matrix

The strain matrix [B] can be written as

$$[B]=[B_i \quad B_j \quad B_k]=\frac{1}{2A}\begin{bmatrix} b_i & 0 & b_j & 0 & b_k & 0 \\ 0 & c_i & 0 & c_j & 0 & c_k \\ c_i & b_i & c_j & b_j & c_k & b_k \end{bmatrix}$$ [shown in Eq. (1−13)]

where $b_i = y_j - y_k, c_i = x_k - x_j (i, j, k)$.

We select the 5th element shown in Fig. 2−2 as an example. In the follows, "Do" means making a circulation, and "Do 100 M =1, 3" means make a circulation from current line to the line numbered 100; the variable M increases from 1 to 3, and the increment is 1.

```
Do 100 M=1, 3              ➡    [make a circulation]
Do 100 N=1, 6
100  B (i, j)= 0.0         ➡    [note: all the terms in [B] matrix equal zero]
     B (1, 1)=CN(j, 2)-CN(k, 2)        ➡    [See [B] matrix above]
     B (1, 3)=CN(k, 2)-CN(i, 2)
     B (1, 5)=CN(i, 2)-CN(j, 2)
     B (2, 2)=Xkj
     B (2, 4)=Xik
     B (2, 6)=Xji
     B (3, 1)=B(2, 2)
     B (3, 2)=B(1, 1)
     B (3, 3)=B(2, 4)
     B (3, 4)=B(1, 3)
     B (3, 5)=B(2, 6)
     B (3, 6)=B(1, 5)
     Do 200 M=1, 3         ➡    [make a circulation, all the terms in [B] matrix/2A]
     Do 200 N=1, 6
200  B(M, N)=0.5*B (M, N)/Area
```

2.3 Calculating [S] stress matrix

The stress matrix can be calculated by [D] times [B]

$$[S]=[D][B]=\frac{E}{2A(1-\nu^2)}\begin{bmatrix}1 & \nu & 0\\ \nu & 1 & 0\\ 0 & 0 & \frac{1-\nu}{2}\end{bmatrix}\begin{bmatrix}b_i & 0 & b_j & 0 & b_k & 0\\ 0 & c_i & 0 & c_j & 0 & c_k\\ c_i & b_i & c_j & b_j & c_k & b_k\end{bmatrix}$$

$$=\frac{E}{2A(1-\nu^2)}\begin{bmatrix}b_i & \nu c_i & b_j & \nu c_j & b_k & \nu c_k\\ \nu b_i & c_i & \nu b_j & c_j & \nu b_k & c_k\\ \frac{1-\nu}{2}c_i & \frac{1-\nu}{2}b_i & \frac{1-\nu}{2}c_j & \frac{1-\nu}{2}b_j & \frac{1-\nu}{2}c_k & \frac{1-\nu}{2}b_k\end{bmatrix}$$

[shown in Eq. (1−18)]

Based on the above equation, the program code is written as

```
      Do 300 M=1, 3
      Do 300 N=1, 3
300   D(M, N)=0.0      ➡    [all the components in [D] matrix equal zero]
      DU=E0/(1 − U0 * U0)
      D(1, 1)=DU
      D(1, 2)=DU * U0
      D(2, 1)=D(1, 2)
      D(2, 2)=DU
      D(3, 3)=0.5 * DU * (1−U0)
      Do 400 M=1, 3
      Do 400 N=1, 6
      S(M, N)=0.0
      Do 400 k=1, 3
400   S(M, N)=S(M, N)+D(M, k) * B(k, N)
```

2.4 Calculating element stiffness matrix $[K]^e$

The element stiffness matrix $[K]^e$ can be obtained directly from Eq. (1−34), i. e.

$$[K]^e=\frac{Et}{4(1-\nu^2)A}\left[\begin{array}{cccccc}
b_ib_i+\frac{1-\nu}{2}c_ic_i & \nu b_ic_i+\frac{1-\nu}{2}c_ib_i & b_ib_j+\frac{1-\nu}{2}c_ic_j & \nu b_ic_j+\frac{1-\nu}{2}c_ib_j & b_ib_k+\frac{1-\nu}{2}c_ic_k & \nu b_ic_k+\frac{1-\nu}{2}c_ib_k\\
\multicolumn{2}{c}{K_{ii}} & \multicolumn{2}{c}{K_{ij}} & \multicolumn{2}{c}{K_{ik}}\\
\nu c_ib_i+\frac{1-\nu}{2}b_ic_i & c_ic_i+\frac{1-\nu}{2}b_ib_i & \nu c_ib_j+\frac{1-\nu}{2}b_ic_j & c_ic_j+\frac{1-\nu}{2}b_ib_j & \nu c_ib_k+\frac{1-\nu}{2}b_ic_k & c_ic_k+\frac{1-\nu}{2}b_ib_k\\
b_ib_j+\frac{1-\nu}{2}c_ic_j & \nu b_jc_i+\frac{1-\nu}{2}c_jb_i & b_jb_j+\frac{1-\nu}{2}c_jc_j & \nu b_jc_j+\frac{1-\nu}{2}c_jb_j & b_jb_k+\frac{1-\nu}{2}c_jc_k & \nu b_jc_k+\frac{1-\nu}{2}c_jb_k\\
\multicolumn{2}{c}{K_{ji}} & \multicolumn{2}{c}{K_{jj}} & \multicolumn{2}{c}{K_{jk}}\\
\nu c_jb_i+\frac{1-\nu}{2}b_jc_i & c_jc_i+\frac{1-\nu}{2}b_jb_i & \nu c_jb_j+\frac{1-\nu}{2}b_jc_j & c_jc_j+\frac{1-\nu}{2}b_jb_j & \nu c_jb_k+\frac{1-\nu}{2}b_jc_k & c_jc_k+\frac{1-\nu}{2}b_jb_k\\
b_kb_i+\frac{1-\nu}{2}c_kc_i & \nu b_kc_i+\frac{1-\nu}{2}c_kb_i & b_kb_j+\frac{1-\nu}{2}c_kc_j & \nu b_kc_j+\frac{1-\nu}{2}c_kb_j & b_kb_k+\frac{1-\nu}{2}c_kc_k & \nu b_kc_k+\frac{1-\nu}{2}c_kb_k\\
\multicolumn{2}{c}{K_{ki}} & \multicolumn{2}{c}{K_{jk}} & \multicolumn{2}{c}{K_{kk}}\\
\nu c_kb_i+\frac{1-\nu}{2}b_kc_i & c_kc_i+\frac{1-\nu}{2}b_kb_i & \nu c_kb_j+\frac{1-\nu}{2}b_kc_j & c_kc_j+\frac{1-\nu}{2}b_kb_j & \nu c_kb_k+\frac{1-\nu}{2}b_kc_k & c_kc_k+\frac{1-\nu}{2}b_kb_k
\end{array}\right]$$

[shown in Eq. (1−34)]

We can also calculate the $[K]^e$ by

$$[K]^e = [B]^T[S]tA = \frac{Et}{4A(1-\nu^2)} \begin{bmatrix} b_i & 0 & c_i \\ 0 & c_i & b_i \\ b_j & 0 & c_j \\ 0 & c_j & b_j \\ b_k & 0 & c_k \\ 0 & c_k & b_k \end{bmatrix} \begin{bmatrix} b_i & \nu c_i & b_j & \nu c_j & b_k & \nu c_k \\ \nu b_i & c_i & \nu b_j & c_j & \nu b_k & c_k \\ \frac{1-\nu}{2}c_i & \frac{1-\nu}{2}b_i & \frac{1-\nu}{2}c_j & \frac{1-\nu}{2}b_j & \frac{1-\nu}{2}c_k & \frac{1-\nu}{2}b_k \end{bmatrix}$$

where $b_i = y_j - y_k, c_i = -x_j + x_k (i, j, k)$.

Based on the above equation, we write the program code as

```
        Do 500 i =1, 6 ⎫
        Do 500 j =1, 6 ⎬all the components in EK equal zero
        EK(i, j) =0.0  ⎭
        Do 500 k =1, 3
        EK(i, j)=EK(i, j)+B(k, i) * S(k, j) * Area * t
500     Continue
        Return
        End
```

We can check the effectiveness of the above code: if i=2, j=4, then from the above code, EK(2, 4)=B(1, 2) * S(1, 4)+B(2, 2) * S(2, 4)+B(3, 2) * S(3, 4).

Chapter 3 Equivalent nodal forces

In finite element method, the interaction between adjacent elements is transferred through the nodes, not the boundaries. In the real situation, the external loadings may not just act on the nodes, but may be along the surfaces or an arbitrary point inside a triangle or on the boundary. Therefore, we have to transform the loads into nodal loads.

3.1 Concentrated load

Supposing there is a concentrated load applied at an arbitrary point (x, y) inside a triangular element, as shown in Fig. 3—1. Because only the loads acting on the nodes can be treated in finite element method, and therefore, we have to transform the concentrated loads equivalently to the three nodes. Now supposing this element has a virtual displacement. The concentrated load and the corresponding virtual displacement can be written as

$$\{P\}=\begin{Bmatrix}P_x\\P_y\end{Bmatrix},\ \{\Delta^*\}=\begin{Bmatrix}u^*\\v^*\end{Bmatrix} \tag{3-1}$$

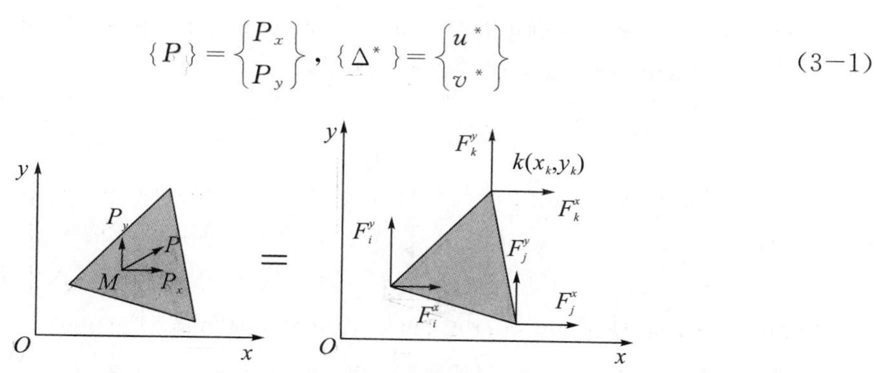

Fig. 3—1 A concentrated load and its equivalent forces

The equivalent nodal forces and the corresponding virtual displacements at point M can be expressed as

$$\{F\}^e = \begin{bmatrix} F_i^x \\ F_i^y \\ F_j^x \\ F_j^y \\ F_k^x \\ F_k^y \end{bmatrix}, \quad \{\delta^*\}^e = \begin{bmatrix} u_i^* \\ v_i^* \\ u_j^* \\ v_j^* \\ u_k^* \\ v_k^* \end{bmatrix} \tag{3-2}$$

The external work done by the equivalent nodal forces are

$$(\{\delta^*\}^e)^{\mathrm{T}}\{F\}^e = [u_i^* \quad v_i^* \quad u_j^* \quad v_j^* \quad u_k^* \quad v_k^*] \begin{bmatrix} F_i^x \\ F_i^y \\ F_j^x \\ F_j^y \\ F_k^x \\ F_k^y \end{bmatrix}$$

$$= F_i^x u_i^* + F_i^y v_i^* + F_j^x u_j^* + F_j^y v_j^* + F_k^x u_k^* + F_k^y v_k^* \tag{3-3}$$

The external work done by the concentrated load P

$$(\{\Delta^*\})^{\mathrm{T}}\{P\} = [u^* \quad v^*]\begin{Bmatrix} P_x \\ P_y \end{Bmatrix} = P_x u^* + P_y v^* \tag{3-4}$$

where $\{\Delta^*\}$ is the virtual displacement of point M. We know that there is a relationship between the displacement of a point and those of the three nodes, which can be expressed as

$$\begin{Bmatrix} u \\ v \end{Bmatrix} = \begin{bmatrix} N_i & 0 & N_j & 0 & N_k & 0 \\ 0 & N_i & 0 & N_j & 0 & N_k \end{bmatrix}\{\delta\}^e = [N]\{\delta\}^e \quad \text{[shown in Eq. (1-10)]}$$

Similarly, the relationship between the virtual displacement of point M and the nodal virtual displacements can be written as

$$\{\Delta^*\} = \begin{Bmatrix} u^* \\ v^* \end{Bmatrix} = [N]\{\delta^*\}^e \tag{3-5}$$

Substituting Eq. (3-5) into Eq. (3-4), the work done by the concentrated load P can be rewritten as

$$(\{\Delta^*\})^{\mathrm{T}}\{P\} = ([N]\{\delta^*\}^e)^{\mathrm{T}}\{P\} = (\{\delta^*\}^e)^{\mathrm{T}}[N]\{P\} \tag{3-6}$$

Suppose the work done by the concentrated load $\{P\}$ is equal to the work done by equivalent nodal forces $\{F\}^e$, then we have

$$(\{\delta^*\})^{\mathrm{T}}\{F\}^e = (\{\Delta^*\})^{\mathrm{T}}\{P\} \tag{3-7}$$

Substituting Eq. (3-6) into Eq. (3-7), we have

$$(\{\delta^*\})^{\mathrm{T}}\{F\}^e = (\{\delta^*\}^e)^{\mathrm{T}}[N]^{\mathrm{T}}\{P\} \tag{3-8}$$

Deleting $(\{\delta^*\})^{\mathrm{T}}$ in both sides of Eq. (3-8), the equivalent nodal force can be expressed as

$$\{F\}^e = [N]^{\mathrm{T}}\{P\} \tag{3-9}$$

More details of Eq. (3-9) is

$$\{F\}^e = \begin{Bmatrix} F_i^x \\ F_i^y \\ F_j^x \\ F_j^y \\ F_k^x \\ F_k^y \end{Bmatrix} = \begin{bmatrix} N_i & 0 \\ 0 & N_i \\ N_j & 0 \\ 0 & N_j \\ N_k & 0 \\ 0 & N_k \end{bmatrix} \begin{Bmatrix} P_x \\ P_y \end{Bmatrix} \tag{3-10}$$

Here the shape functions are

$$N_i = \frac{1}{2A}(a_i + b_i x + c_i y) \quad (i,j,k) \quad [\text{shown in Eq. (1-9)}]$$

and the coefficients are

$$\begin{cases} a_i = x_j y_k - x_k y_j \\ b_i = y_j - y_k \\ c_i = -x_j + x_k \end{cases} \quad (i,j,k)$$

3.2 Body force

We treat the body force as a concentrated force, i. e. the body force of the whole element, such as the gravity force, is considered as a concentrated force acting on the gravity center (centroid), as shown in Fig. 3-2. The gravity force only has one component in y direction and can be expressed as

$$\{P\} = \begin{Bmatrix} 0 \\ -\rho g t A \end{Bmatrix} \tag{3-11}$$

where ρ is density of the material, t is thickness, A is the triangle area which can be calculated from Eq. (1-5).

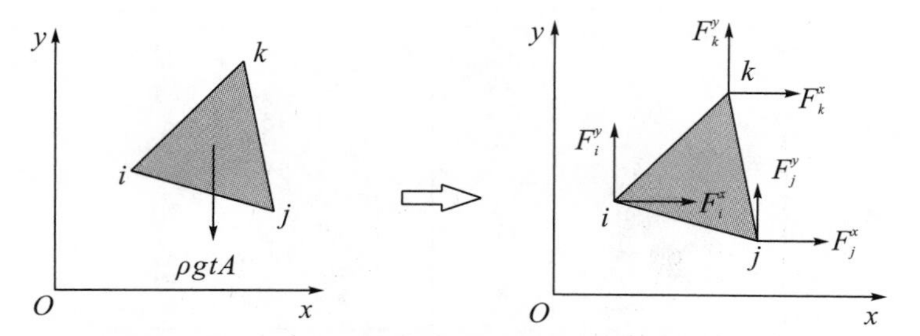

Fig. 3-2 The body force and equivalent nodal forces

According to the result of concentrate load, one can have

$$\{F\}^e=\begin{Bmatrix}F_i^x\\F_i^y\\F_j^x\\F_j^y\\F_k^x\\F_k^y\end{Bmatrix}=\begin{bmatrix}N_i&0\\0&N_i\\N_j&0\\0&N_j\\N_k&0\\0&N_k\end{bmatrix}\begin{Bmatrix}P_x\\P_y\end{Bmatrix}=\begin{bmatrix}N_i&0\\0&N_i\\N_j&0\\0&N_j\\N_k&0\\0&N_k\end{bmatrix}\begin{Bmatrix}0\\-\rho gtA\end{Bmatrix}\tag{3-12}$$

The coordinates of the centroid can be expressed as $x=\frac{1}{3}(x_i+x_j+x_k)$, $y=\frac{1}{3}(y_i+y_j+y_k)$. Substituting the centroid coordinates x and y into $N_i=\frac{1}{2A}(a_i+b_ix+c_iy)$, we get $N_i=N_j=N_k=\frac{1}{3}$. From Eq. (3−12), one can have

$$\{F\}^e=\begin{Bmatrix}F_i^x\\F_i^y\\F_j^x\\F_j^y\\F_k^x\\F_k^y\end{Bmatrix}=\begin{bmatrix}N_i&0\\0&N_i\\N_j&0\\0&N_j\\N_k&0\\0&N_k\end{bmatrix}\begin{Bmatrix}0\\-\rho gtA\end{Bmatrix}=\begin{Bmatrix}0\\-N_i\rho gtA\\0\\-N_j\rho gtA\\0\\-N_k\rho gtA\end{Bmatrix}=\begin{Bmatrix}0\\-\frac{1}{3}\rho gtA\\0\\-\frac{1}{3}\rho gtA\\0\\-\frac{1}{3}\rho gtA\end{Bmatrix}\tag{3-13}$$

From Eq. (3−13), we can find that the body force is equally assigned to the three nodes.

In addition, we can also calculate the equivalent node forces using integration method. We take a small body from the triangle, the load per unit can be written as

$$\{p\}=\begin{Bmatrix}X\\Y\end{Bmatrix}\tag{3-14}$$

For gravity, $X=0$, $Y=-\rho g$.

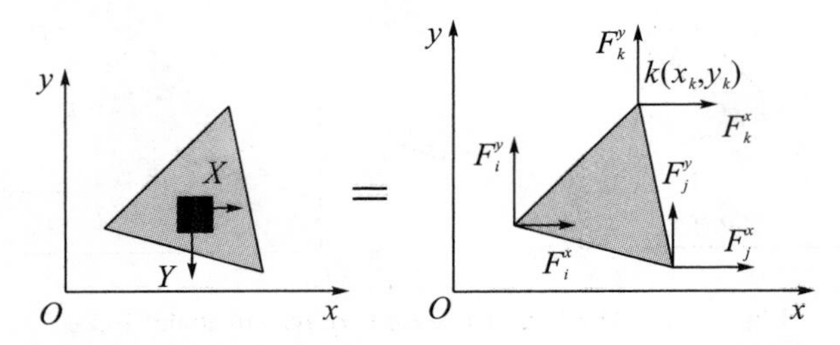

Fig. 3−3 Transforming the uniform body force to nodal forces

The volume of this small body is $t\mathrm{d}x\mathrm{d}y$, and the body force is $\begin{Bmatrix}X\\Y\end{Bmatrix}t\mathrm{d}x\mathrm{d}y$. Substituting the body force into Eq. (3−12), one can have the equivalent forces of this small body

$$\{F\}^e = \begin{Bmatrix} F_i^x \\ F_i^y \\ F_j^x \\ F_j^y \\ F_k^x \\ F_k^y \end{Bmatrix} = \begin{bmatrix} N_i & 0 \\ 0 & N_i \\ N_j & 0 \\ 0 & N_j \\ N_k & 0 \\ 0 & N_k \end{bmatrix} \begin{Bmatrix} X \\ Y \end{Bmatrix} t\,\mathrm{d}x\,\mathrm{d}y$$

For the whole element, the equivalent forces of this element can be written as

$$\{F\}^e = \begin{Bmatrix} F_i^x \\ F_i^y \\ F_j^x \\ F_j^y \\ F_k^x \\ F_k^y \end{Bmatrix} = \iint \begin{bmatrix} N_i & 0 \\ 0 & N_i \\ N_j & 0 \\ 0 & N_j \\ N_k & 0 \\ 0 & N_k \end{bmatrix} \begin{Bmatrix} X \\ Y \end{Bmatrix} t\,\mathrm{d}x\,\mathrm{d}y = \begin{Bmatrix} 0 \\ -\rho g t \iint N_i\,\mathrm{d}x\,\mathrm{d}y \\ 0 \\ -\rho g t \iint N_j\,\mathrm{d}x\,\mathrm{d}y \\ 0 \\ -\rho g t \iint N_k\,\mathrm{d}x\,\mathrm{d}y \end{Bmatrix} = \begin{Bmatrix} 0 \\ -\frac{1}{3}\rho g t A \\ 0 \\ -\frac{1}{3}\rho g t A \\ 0 \\ -\frac{1}{3}\rho g t A \end{Bmatrix} \tag{3-15}$$

where $\iint N_i\,\mathrm{d}x\,\mathrm{d}y = \iint N_j\,\mathrm{d}x\,\mathrm{d}y = \iint N_k\,\mathrm{d}x\,\mathrm{d}y = \frac{1}{3}$.

3.3 Distributed force

Supposing the distributed force is acting on the boundary of an element, and in a unit area, the distributed stress is

$$\{q\} = \begin{Bmatrix} \overline{X} \\ \overline{Y} \end{Bmatrix} \tag{3-16}$$

The distributed forces can also be considered as a concentrated force. For the distributed force shown in Fig. 3-4, the total force acting on the line ki is qlt, acting on the center of line ki. According to the result of concentrated load, we have

$$\{F\}^e = \begin{Bmatrix} F_i^x \\ F_i^y \\ F_j^x \\ F_j^y \\ F_k^x \\ F_k^y \end{Bmatrix} = \begin{bmatrix} N_i & 0 \\ 0 & N_i \\ N_j & 0 \\ 0 & N_j \\ N_k & 0 \\ 0 & N_k \end{bmatrix} \begin{Bmatrix} \overline{X}lt \\ \overline{Y}lt \end{Bmatrix} \tag{3-17}$$

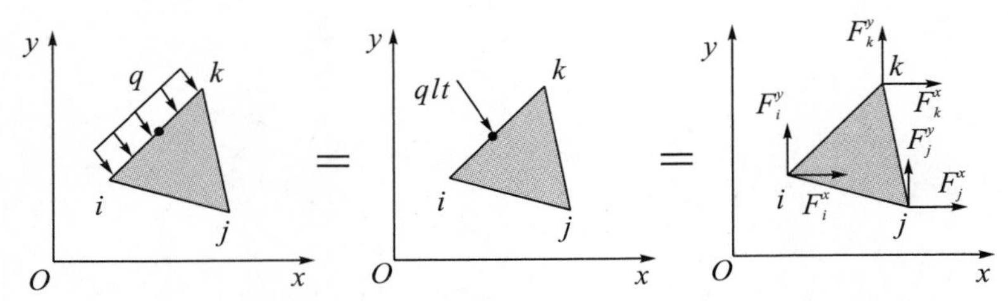

Fig. 3−4 Transforming distribute forces in nodal forces

The coordinates of the center of line ki are $x = \frac{x_i + x_k}{2}, y = \frac{y_i + y_k}{2}$. Substituting them into the shape function $N_i = \frac{1}{2A}(a_i + b_i x + c_i y)$, and combining with the coefficients $a_i = x_j y_k - x_k y_j, b_i = y_j - y_k, c_i = -x_j + x_k$, we get $N_i = 0.5, N_j = 0, N_k = 0.5$. Substituting them into Eq. (3−17), we have

$$\{F\}^e = \begin{Bmatrix} F_i^x \\ F_i^y \\ F_j^x \\ F_j^y \\ F_k^x \\ F_k^y \end{Bmatrix} = \begin{bmatrix} N_i & 0 \\ 0 & N_i \\ N_j & 0 \\ 0 & N_j \\ N_k & 0 \\ 0 & N_k \end{bmatrix} \begin{Bmatrix} \overline{X}lt \\ \overline{Y}lt \end{Bmatrix} = \begin{Bmatrix} \frac{1}{2}lt\overline{X} \\ \frac{1}{2}lt\overline{Y} \\ 0 \\ 0 \\ \frac{1}{2}lt\overline{X} \\ \frac{1}{2}lt\overline{Y} \end{Bmatrix} \tag{3−18}$$

If the distributed force is parallel to x-axis, as shown in Fig. 3−5, the equivalent nodal force can be written

$$\{F\}^e = \begin{Bmatrix} F_i^x \\ F_i^y \\ F_j^x \\ F_j^y \\ F_k^x \\ F_k^y \end{Bmatrix} = \begin{bmatrix} N_i & 0 \\ 0 & N_i \\ N_j & 0 \\ 0 & N_j \\ N_k & 0 \\ 0 & N_k \end{bmatrix} \begin{Bmatrix} \overline{X}lt \\ 0 \end{Bmatrix} = \begin{Bmatrix} \frac{1}{2}lt\overline{X} \\ 0 \\ 0 \\ 0 \\ \frac{1}{2}lt\overline{X} \\ 0 \end{Bmatrix} \tag{3−19}$$

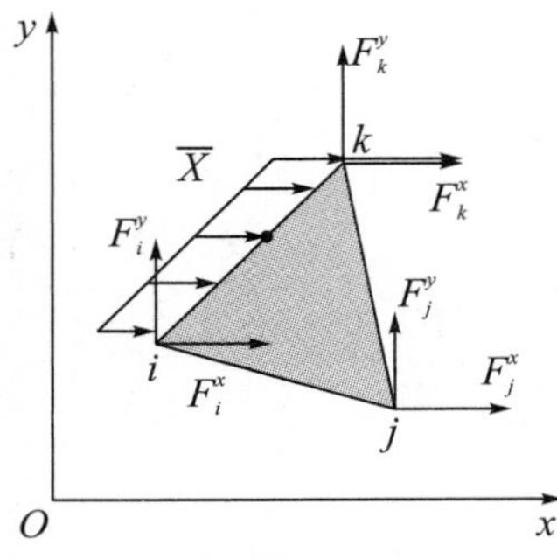

Fig. 3−5 The distributed force paralleling to *x*-axis

If the force is distributed triangularly as shown in Fig. 3-6, we can get $N_i = \frac{1}{3}$, $N_j = 0, N_k = \frac{2}{3}$, and the equivalent nodal force can be written as

$$\{F\}^e = \begin{Bmatrix} F_i^x \\ F_i^y \\ F_j^x \\ F_j^y \\ F_k^x \\ F_k^y \end{Bmatrix} = \begin{Bmatrix} N_i & 0 \\ 0 & N_i \\ N_j & 0 \\ 0 & N_j \\ N_k & 0 \\ 0 & N_k \end{Bmatrix} \begin{Bmatrix} \frac{1}{2}\overline{X}lt \\ 0 \end{Bmatrix} = \begin{Bmatrix} \frac{1}{3}\left(\frac{1}{2}lt\overline{X}\right) \\ 0 \\ 0 \\ 0 \\ \frac{2}{3}\left(\frac{1}{2}lt\overline{X}\right) \\ 0 \end{Bmatrix} \tag{3-20}$$

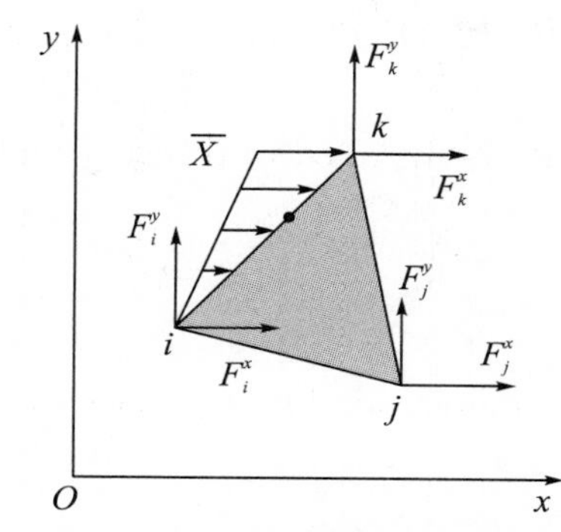

Fig. 3-6 The force distributed triangularly

From the above study, we can find that the calculation method of the equivalent forces for concentrated loads is very important since all the other kinds of loads, such as the body loads and the distributed loads, can be considered as a concentrated load. Therefore, the students should completely grasp the following equation in this study.

$$\{F\}^e = \begin{Bmatrix} F_i^x \\ F_i^y \\ F_j^x \\ F_j^y \\ F_k^x \\ F_k^y \end{Bmatrix} = \begin{Bmatrix} N_i & 0 \\ 0 & N_i \\ N_j & 0 \\ 0 & N_j \\ N_k & 0 \\ 0 & N_k \end{Bmatrix} \begin{Bmatrix} P_x \\ P_y \end{Bmatrix} \quad \text{[shown in Eq. (3-10)]}$$

3.4 Subroutine for body load

Loads include concentrated load, body load and distributed load. Here we select a body load as an example to illustrate. The subroutine name is load, and the parameters are transmitted through the subsequent bracket.

```
    Subroutine Load(NNE, MNE, CN, MNN, NN2, P, Gam)
    Dimension NNE(MNE, 3), CN(MNN, 2), P(NN2)
    If(Gam. LE. 0.0)go to 200   ➡  [if γ≤0, go to 200]
    Do 100 NE=1, MNE   ➡  [make a circulation]
    i=NGN(NE, 1) ┐
    j=NGN(NE, 2) ├ [three numbers of an element]
    k=NGN(NE, 3) ┘
    Xji=CN(j, 1)-CN(i, 1)
    Xik=CN(i, 1)-CN(k, 1)
    Xkj=CN(k, 1)-CN(j, 1)
    Area=0.5 * (CN(k, 2) * Xji+CN(j, 2) * Xik+CN(i, 2) * Xkj)  [triangle aera]
    PE=-Gam * Area * t/3.0  ➡  [PE=- γAt/3, γ=ρg]
    P(2 * i)=P(2 * i)+PE ┐
    P(2 * j)=P(2 * j)+PE ├PE is installed to the i, j and k nodal loads in y-axis
200 P(2 * k)=P(2 * k)+PE ┘
    End
```

NNE=Node Number of Element
CN=Coordinates of Nodes
MNE=Maximum Number of Element
MNN=Maximum Number of Nodes
P=Global load vector
NN2=2 * MNN
GAM=ρg

The students need to write more code for the concentrated loads and the distributed loads.

Assignments:

The nodes of triangular element are i, j and k, and a linear distributed load is applied on the edge jk as shown in the following figure. Calculate the three-node loads.

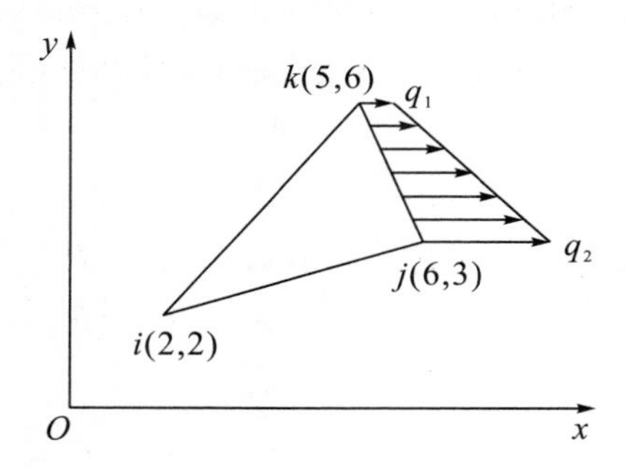

Chapter 4 Global stiffness matrix

Usually there are many elements which are connected through nodes, and around a node, there may be several elements, such as the node 4 in Fig. 4−1.

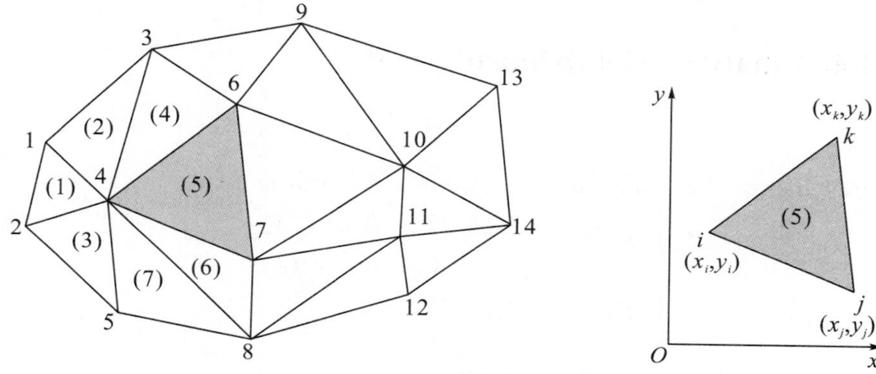

Fig. 4−1 Meshing of a domain

4.1 Global stiffness matrix and its property

For a single element, based on the principle of virtual displacement, we have

$$(\{\delta^*\})^{\mathrm{T}}\{F\}^e = \iint_A \{\varepsilon^*\}\{\sigma\}t\,\mathrm{d}x\,\mathrm{d}y \quad [\text{shown in Eq. (1−25)}]$$

and we get the important equilibrium equation

$$\begin{bmatrix} F_i^x \\ F_i^y \\ F_j^x \\ F_j^y \\ F_k^x \\ F_k^y \end{bmatrix} = \frac{Et}{4(1-\nu^2)A}\begin{bmatrix} b_ib_i+\frac{1-\nu}{2}c_ic_i & \nu b_ic_i+\frac{1-\nu}{2}c_ib_i & b_ib_j+\frac{1-\nu}{2}c_ic_j & \nu b_ic_j+\frac{1-\nu}{2}c_ib_j & b_ib_k+\frac{1-\nu}{2}c_ic_k & \nu b_ic_k+\frac{1-\nu}{2}c_ib_k \\ \nu c_ib_i+\frac{1-\nu}{2}b_ic_i & c_ic_i+\frac{1-\nu}{2}b_ib_i & \nu c_ib_j+\frac{1-\nu}{2}b_ic_j & c_ic_j+\frac{1-\nu}{2}b_ib_j & \nu c_ib_k+\frac{1-\nu}{2}b_ic_k & c_ic_k+\frac{1-\nu}{2}b_ib_k \\ b_ib_j+\frac{1-\nu}{2}c_ic_j & \nu b_jc_i+\frac{1-\nu}{2}c_jb_i & b_jb_j+\frac{1-\nu}{2}c_jc_j & \nu b_jc_j+\frac{1-\nu}{2}c_jb_j & b_jb_k+\frac{1-\nu}{2}c_jc_k & \nu b_jc_k+\frac{1-\nu}{2}c_jb_k \\ \nu c_jb_i+\frac{1-\nu}{2}b_jc_i & c_jc_i+\frac{1-\nu}{2}b_jb_i & \nu c_jb_j+\frac{1-\nu}{2}b_jc_j & c_jc_j+\frac{1-\nu}{2}b_jb_j & \nu c_jb_k+\frac{1-\nu}{2}b_jc_k & c_jc_k+\frac{1-\nu}{2}b_jb_k \\ b_kb_i+\frac{1-\nu}{2}c_kc_i & \nu b_kc_i+\frac{1-\nu}{2}c_kb_i & b_kb_j+\frac{1-\nu}{2}c_kc_j & \nu b_kc_j+\frac{1-\nu}{2}c_kb_j & b_kb_k+\frac{1-\nu}{2}c_kc_k & \nu b_kc_k+\frac{1-\nu}{2}c_kb_k \\ \nu c_kb_i+\frac{1-\nu}{2}b_kc_i & c_kc_i+\frac{1-\nu}{2}b_kb_i & \nu c_kb_j+\frac{1-\nu}{2}b_kc_j & c_kc_j+\frac{1-\nu}{2}b_kb_j & \nu c_kb_k+\frac{1-\nu}{2}b_kc_k & c_kc_k+\frac{1-\nu}{2}b_kb_k \end{bmatrix}\begin{bmatrix} u_i \\ v_i \\ u_j \\ v_j \\ u_k \\ v_k \end{bmatrix}$$

(block labels: K_{ii}, K_{ij}, K_{ik} / K_{ji}, K_{jj}, K_{jk} / K_{ki}, K_{kj}, K_{kk})

[shown in Eq. (1−36)]

or

$$\{F\}^e = [K]^e \{\delta\}^e \qquad \text{[shown in Eq. (1-31)]}$$

For each element, the above equation holds, and then for all the elements, by accumulation method we have

$$\sum \{F\}^e = \sum [K]^e \{\delta\}^e \tag{4-1}$$

Eq. (4-1) can be simplified as

$$\{F\} = [K]\{\delta\} \tag{4-2}$$

where $[K]$ is global stiffness matrix, $\{F\}$ is nodal force vector, $\{\delta\}$ is nodal displacement vector.

4.2 Global matrix establishment

For an element, the equilibrium equation can be written as

$$\begin{Bmatrix} \{F_i\} \\ \{F_j\} \\ \{F_k\} \end{Bmatrix} = \begin{bmatrix} K_{ii} & K_{ij} & K_{ik} \\ K_{ji} & K_{jj} & K_{jk} \\ K_{ki} & K_{kj} & K_{kk} \end{bmatrix} \begin{Bmatrix} \{\delta_i\} \\ \{\delta_j\} \\ \{\delta_k\} \end{Bmatrix} \tag{4-3}$$

where

$$[K_{rs}] = \frac{Et}{4(1-\nu^2)A} \begin{bmatrix} b_r b_s + \frac{1-\nu}{2} c_r c_s & \nu b_r c_s + \frac{1-\nu}{2} c_r b_s \\ \nu c_r b_s + \frac{1-\nu}{2} b_r c_s & c_r c_s + \frac{1-\nu}{2} b_r b_s \end{bmatrix} (r = i, j, k; s = i, j, k)$$

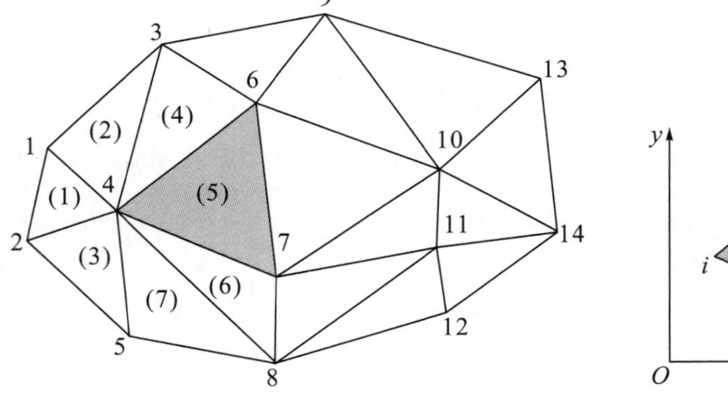

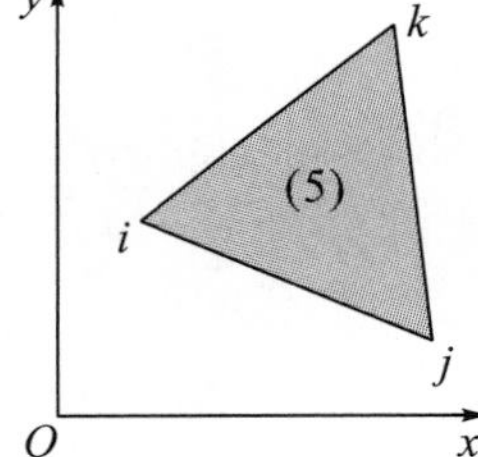

Fig. 4-2 The element (5)

For each element, all the i, j and k have specific numbers. We use the real element numbers to replace the i, j, k numbers. For element (5) shown in Fig. 4-2, $i = 4, j = 7, k = 6$, and then $K_{ii} \rightarrow K_{44}, K_{ij} \rightarrow K_{47}$, and $K_{kj} \rightarrow K_{67}$. The equilibrium equation of element (5) can be written as

$$\begin{Bmatrix}\{F_4\}\\\{F_7\}\\\{F_6\}\end{Bmatrix}=\begin{bmatrix}K_{44} & K_{47} & K_{46}\\K_{74} & K_{77} & K_{76}\\K_{64} & K_{67} & K_{66}\end{bmatrix}\begin{Bmatrix}\{\delta_4\}\\\{\delta_7\}\\\{\delta_6\}\end{Bmatrix}\tag{4—4}$$

Next we will select a domain consisting of three elements and five nodes as shown in Fig. 4—3 to analyze. For each point, we will establish an equilibrium equation.

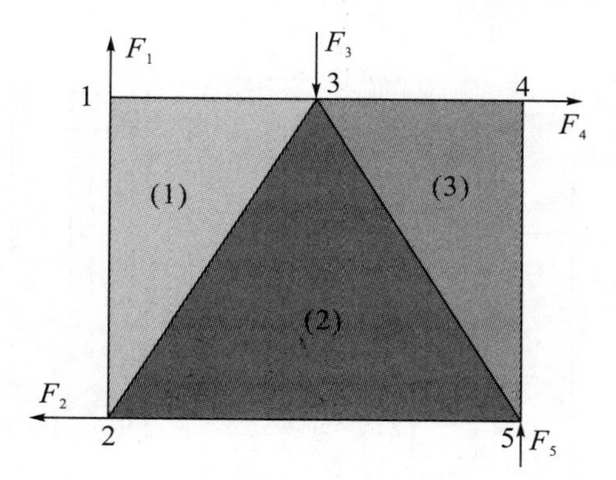

Fig. 4—3 Three elements and five nodes

For element (1), $i=2$, $j=3$ and $k=1$, and the corresponding equilibrium equation is

$$\begin{bmatrix}K_{22} & K_{23} & K_{21}\\K_{32} & K_{33} & K_{31}\\K_{12} & K_{13} & K_{11}\end{bmatrix}\begin{Bmatrix}\{\delta_2\}\\\{\delta_3\}\\\{\delta_1\}\end{Bmatrix}=\begin{Bmatrix}\{F_2\}\\\{F_3\}\\\{F_1\}\end{Bmatrix}\tag{4—5}$$

For element (2), $i=5$, $j=3$ and $k=2$, and the corresponding equilibrium equation is

$$\begin{bmatrix}K_{55} & K_{53} & K_{52}\\K_{35} & K_{33} & K_{32}\\K_{25} & K_{23} & K_{22}\end{bmatrix}\begin{Bmatrix}\{\delta_5\}\\\{\delta_3\}\\\{\delta_2\}\end{Bmatrix}=\begin{Bmatrix}\{F_5\}\\\{F_3\}\\\{F_2\}\end{Bmatrix}\tag{4—6}$$

For element (3), $i=5$, $j=4$ and $k=3$, and the corresponding equilibrium equation is

$$\begin{bmatrix}K_{55} & K_{54} & K_{53}\\K_{45} & K_{44} & K_{43}\\K_{35} & K_{34} & K_{33}\end{bmatrix}\begin{Bmatrix}\{\delta_5\}\\\{\delta_4\}\\\{\delta_3\}\end{Bmatrix}=\begin{Bmatrix}\{F_5\}\\\{F_4\}\\\{F_3\}\end{Bmatrix}\tag{4—7}$$

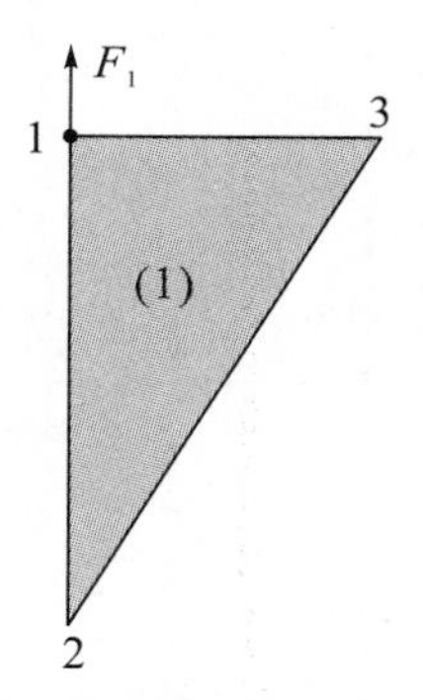

Fig. 4—4 Force balance of on node 1

For node 1, the element 1 has an action on node 1 to resist the tensile force F_1, as shown in Fig. 4—4. From Eq. (4—5), the force acting on node 1 by element (1) can be written as

$$K_{12}^1\delta_2 + K_{13}^1\delta_3 + K_{11}^1\delta_1 = \{F_1\} \tag{4—8}$$

where the superscript 1 denotes element 1. We write Eq. (4 — 8) into global stiffness matrix

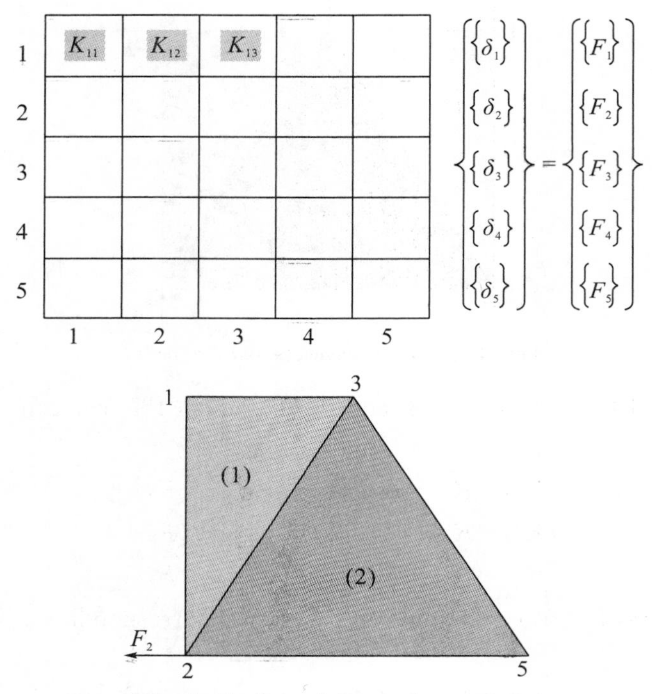

Fig. 4—5 Force balance of on node 2

For node 2, both element (1) and element (2) have action on node 2 to resist the tensile force F_2. From Eqs. (4—5) and (4—6), the forces applied on node 2 from element (1) and element (2) are

$$\begin{cases} K_{22}^1\delta_2 + K_{23}^1\delta_3 + K_{21}^1\delta_1 = \{F_2\}^1 \\ K_{25}^2\delta_5 + K_{23}^2\delta_3 + K_{22}^2\delta_2 = \{F_2\}^2 \end{cases} \tag{4—9}$$

The sum of forces $\{F_2\}^1$ and $\{F_2\}^2$ should be equal to $\{F_2\}$, i. e.

$$K_{22}^1\delta_2 + K_{23}^1\delta_3 + K_{21}^1\delta_1 + K_{25}^2\delta_5 + K_{23}^2\delta_3 + K_{22}^2\delta_2 = \{F_2\}^1 + \{F_2\}^2 = \{F_2\}$$

and the equilibrium equation can be written as

$$K_{21}^1\delta_1 + (K_{22}^1 + K_{22}^2)\delta_2 + (K_{23}^1 + K_{23}^2)\delta_3 + K_{25}^2\delta_5 = \{F_2\} \tag{4—10}$$

Similarly, we write Eq. (4—10) into the global equilibrium equation as

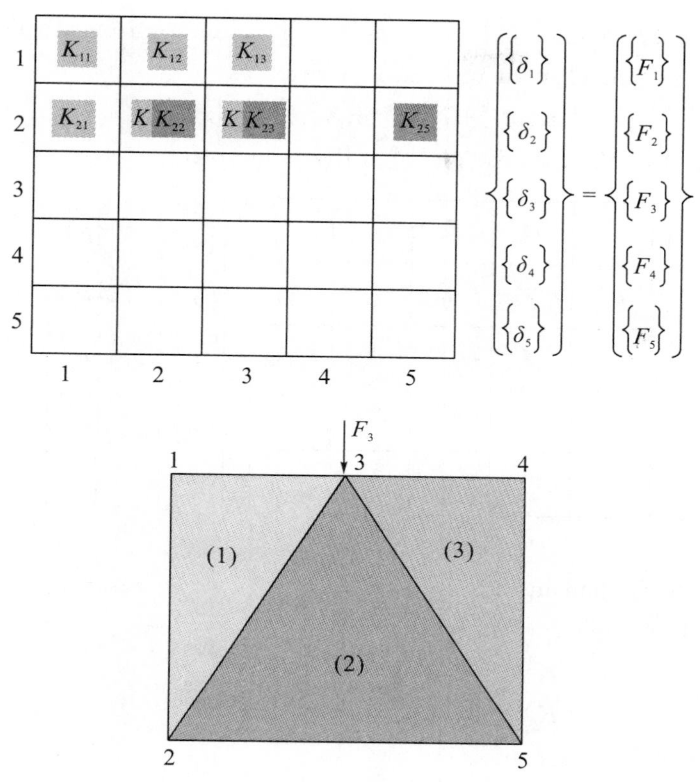

Fig. 4—6 Force balance on node 3

For node 3, all the three elements have contribution to resist the force F_3, and according to Eqs. (4—5), (4—6) and (4—7), the forces acting on node 3 by elements (1), (2) and (3) can be expressed as

$$\begin{cases} K^1_{32}\delta_2 + K^1_{33}\delta_3 + K^1_{31}\delta_1 = \{F_3\}^1 \\ K^2_{35}\delta_5 + K^2_{33}\delta_3 + K^2_{32}\delta_2 = \{F_3\}^2 \\ K^3_{35}\delta_5 + K^3_{34}\delta_4 + K^3_{33}\delta_3 = \{F_3\}^3 \end{cases} \tag{4—11}$$

The sum of forces $\{F_3\}^1$, $\{F_3\}^2$ and $\{F_3\}^3$ should be equal to $\{F_3\}$, i. e.

$$K^1_{31}\delta_1 + (K^1_{32} + K^2_{32})\delta_2 + (K^1_{33} + K^2_{33} + K^3_{33})\delta_3 + K^2_{34}\delta_4 + (K^2_{35} + K^3_{35})\delta_5 = \{F_3\} \tag{4—12}$$

We write Eq. (4—12) into the global stiffness matrix

	1	2	3	4	5
1	K_{11}	K_{12}	K_{13}		
2	K_{21}	K_2K_{22}	K_2K_{23}		K_{25}
3	K_{31}	K_3K_{32}	$KK\,K_{33}$	K_{34}	K_3K_{35}
4					
5					

$$\begin{Bmatrix} \{\delta_1\} \\ \{\delta_2\} \\ \{\delta_3\} \\ \{\delta_4\} \\ \{\delta_5\} \end{Bmatrix} = \begin{Bmatrix} \{F_1\} \\ \{F_2\} \\ \{F_3\} \\ \{F_4\} \\ \{F_5\} \end{Bmatrix}$$

For node 4, only element (3) has action on it, and according to Eq. (4−7), the force acting on node 4 by element (3) can be written as

$$K_{45}^{3}\delta_5+K_{44}^{3}\delta_4+K_{43}^{3}\delta_3=\{F_4\} \tag{4−13}$$

We write Eq. (4−13) into the global stiffness matrix

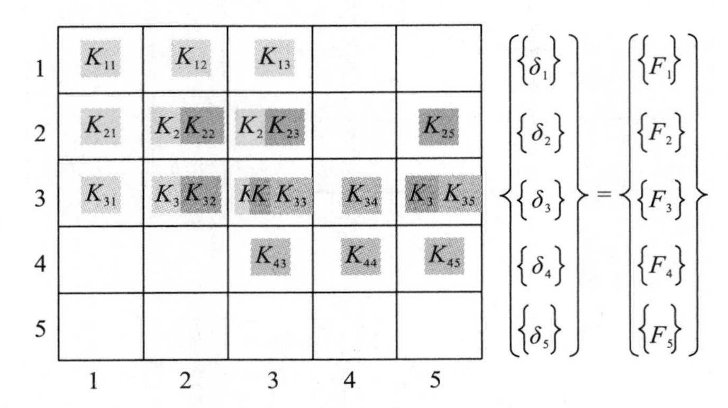

For node 5, both elements (2) and (3) are involved in resisting the force F_5, and according Eqs. (4−6) and (4−7), we can have

$$\begin{aligned}K_{55}^{2}\delta_5+K_{53}^{2}\delta_3+K_{52}^{2}\delta_2&=\{F_5\}^2\\K_{55}^{3}\delta_5+K_{54}^{3}\delta_4+K_{53}^{3}\delta_3&=\{F_5\}^3\end{aligned} \tag{4−14}$$

and the simple equilibrium equation is

$$K_{52}^{2}\delta_2+(K_{53}^{2}+K_{53}^{3})\delta_3+K_{54}^{3}\delta_4+(K_{55}^{2}+K_{55}^{3})\delta_5=\{F_5\} \tag{4−15}$$

We write Eq. (4−15) into the global stiffness matrix as

	1	2	3	4	5
1	K_{11}	K_{12}	K_{13}		
2	K_{21}	K_1K_{22}	K_2K_{23}		K_{25}
3	K_{31}	K_3K_{32}	kKK_{33}	K_{34}	$K_3\ K_{35}$
4			K_{43}	K_{44}	K_{45}
5		K_{52}	KK_{53}	K_{54}	KK_{55}

$$\begin{Bmatrix}\{\delta_1\}\\\{\delta_2\}\\\{\delta_3\}\\\{\delta_4\}\\\{\delta_5\}\end{Bmatrix}=\begin{Bmatrix}\{F_1\}\\\{F_2\}\\\{F_3\}\\\{F_4\}\\\{F_5\}\end{Bmatrix} \tag{4−16}$$

It should be noted that for each load and each displacement in Eq. (4−16), there are two components, i. e.

$$\{\delta\}=\begin{Bmatrix}\delta^x\\\delta^y\end{Bmatrix},\qquad \{F\}=\begin{Bmatrix}F^x\\F^y\end{Bmatrix}$$

and in the global matrix, each term contains four factors. The detailed global stiffness matrix can be written as

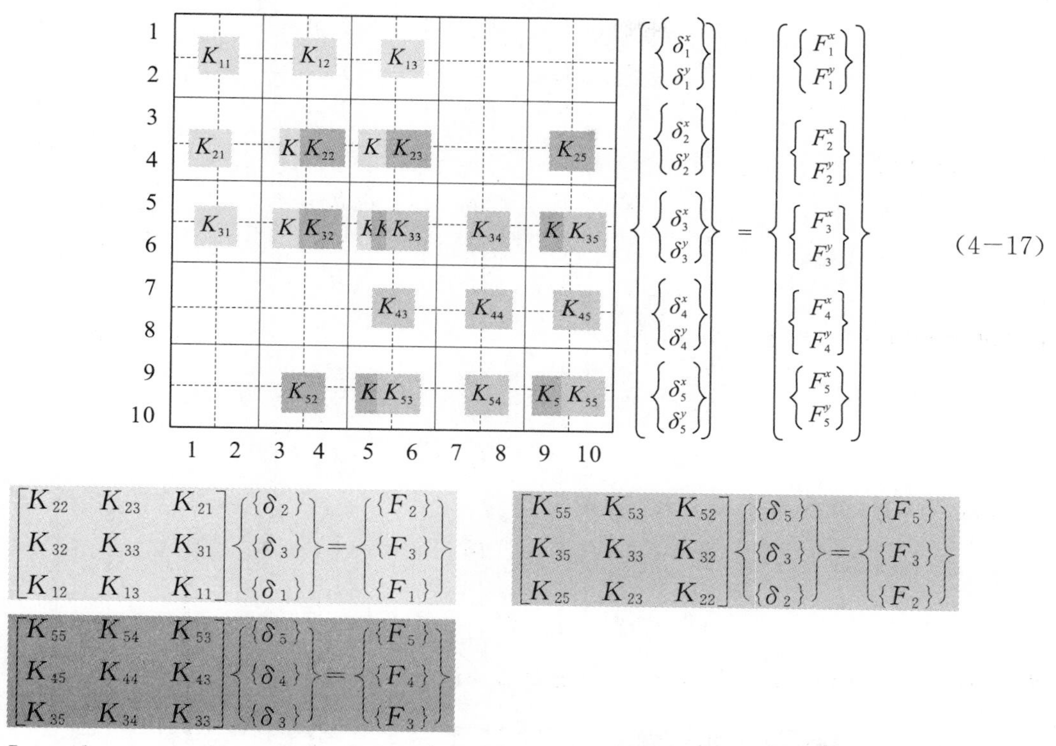

In order to analyze the relationship between the global stiffness matrix and the element stiffness matrix, the three element stiffness matrices are shown above. We can find that the global stiffness matrix consists of the three element stiffness matrices. All the terms in an element stiffness matrix should be installed into the global stiffness matrix according to the subscript numbers.

For a domain divided by many elements as shown in Fig. 4−7, we arbitrarily select several elements, such as element (1), (2), (3) and (5) to practice. According to Eq. (4−4), the stiffness matrices of the four elements can be written as

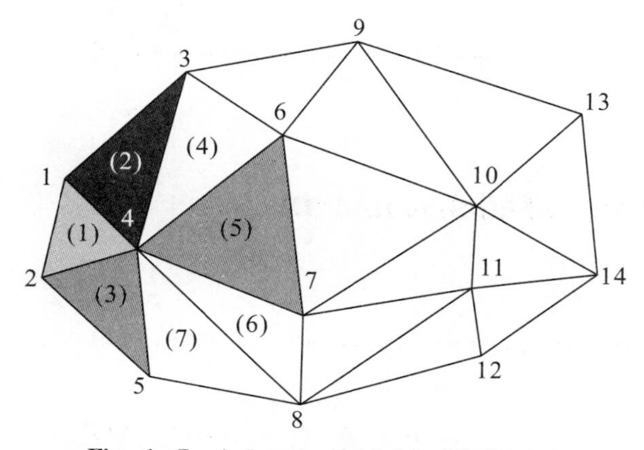

Fig. 4−7 A domain divided by 17 elements

For element (1)($i=2, j=4, k=1$)

$$\begin{bmatrix} K_{22} & K_{24} & K_{21} \\ K_{42} & K_{44} & K_{41} \\ K_{12} & K_{14} & K_{11} \end{bmatrix} \begin{Bmatrix} \{\delta_2\} \\ \{\delta_4\} \\ \{\delta_1\} \end{Bmatrix} = \begin{Bmatrix} \{F_2\} \\ \{F_4\} \\ \{F_1\} \end{Bmatrix}$$

For element (2)($i = 1, j = 4, k = 3$)

$$\begin{bmatrix} K_{11} & K_{14} & K_{13} \\ K_{41} & K_{44} & K_{43} \\ K_{31} & K_{34} & K_{33} \end{bmatrix} \begin{Bmatrix} \{\delta_1\} \\ \{\delta_4\} \\ \{\delta_3\} \end{Bmatrix} = \begin{Bmatrix} \{F_1\} \\ \{F_4\} \\ \{F_3\} \end{Bmatrix}$$

For element (3)($i = 5, j = 4, k = 2$)

$$\begin{bmatrix} K_{55} & K_{54} & K_{52} \\ K_{45} & K_{44} & K_{42} \\ K_{25} & K_{24} & K_{22} \end{bmatrix} \begin{Bmatrix} \{\delta_5\} \\ \{\delta_4\} \\ \{\delta_2\} \end{Bmatrix} = \begin{Bmatrix} \{F_5\} \\ \{F_4\} \\ \{F_2\} \end{Bmatrix}$$

For element (5)($i = 4, j = 7, k = 6$)

$$\begin{bmatrix} K_{44} & K_{47} & K_{46} \\ K_{74} & K_{77} & K_{76} \\ K_{64} & K_{67} & K_{66} \end{bmatrix} \begin{Bmatrix} \{\delta_4\} \\ \{\delta_7\} \\ \{\delta_6\} \end{Bmatrix} = \begin{Bmatrix} \{F_4\} \\ \{F_7\} \\ \{F_6\} \end{Bmatrix}$$

We write all the four matrices into the global stiffness matrix as

$K_{11\,11}$	K_{12}	K_{13}	$K_{14\,14}$						
K_{21}	$K_{22\,22}$		$K_{24\,4}$	K_{25}					
K_{31}		K_{33}	K_{34}						
$K_{41\,41}$	$K_{42\,42}$	K_{43}	$K_{44\,4\,4\,4}$	K_{45}	K_{46}	K_{47}			
	K_{52}		K_{54}	K_{55}					
			K_{64}		K_{66}	K_{67}			
			K_{74}		K_{76}	K_{77}			

$$\begin{Bmatrix} \{\delta_1\} \\ \{\delta_2\} \\ \{\delta_3\} \\ \{\delta_4\} \\ \{\delta_5\} \\ \{\delta_6\} \\ \{\delta_7\} \\ \vdots \\ \{\delta_{14}\} \end{Bmatrix} = \begin{Bmatrix} \{F_1\} \\ \{F_2\} \\ \{F_3\} \\ \{F_4\} \\ \{F_5\} \\ \{F_6\} \\ \{F_7\} \\ \vdots \\ \{F_{14}\} \end{Bmatrix}$$

4.3 The properties of global matrix

(1) From the above study, we can find that the dimension of global matrix should be $2n \times 2n$, where n is the total node number.

(2) The node number i, j and k of an element should be in an anticlockwise sequence, but the initial point i can be selected arbitrarily, which will not affect the global stiffness matrix.

For element (1) in Fig. 4-7, according to the following equation

$$\begin{Bmatrix} \{F_i\} \\ \{F_j\} \\ \{F_k\} \end{Bmatrix} = \begin{bmatrix} K_{ii} & K_{ij} & K_{ik} \\ K_{ji} & K_{jj} & K_{jk} \\ K_{ki} & K_{kj} & K_{kk} \end{bmatrix} \begin{Bmatrix} \{\delta_i\} \\ \{\delta_j\} \\ \{\delta_k\} \end{Bmatrix} \quad \text{[shown in Eq. (1-37)]}$$

If the $i-j-k$ is 2—4—1, we can have

$$\begin{bmatrix} K_{22} & K_{24} & K_{21} \\ K_{42} & K_{44} & K_{41} \\ K_{12} & K_{14} & K_{11} \end{bmatrix} \begin{Bmatrix} \{\delta_2\} \\ \{\delta_4\} \\ \{\delta_1\} \end{Bmatrix} = \begin{Bmatrix} \{F_2\} \\ \{F_4\} \\ \{F_1\} \end{Bmatrix} \tag{4-18}$$

If the $i-j-k$ sequence is 4—1—2, the equation can be written as

$$\begin{bmatrix} K_{44} & K_{41} & K_{42} \\ K_{14} & K_{11} & K_{12} \\ K_{24} & K_{21} & K_{22} \end{bmatrix} \begin{Bmatrix} \{\delta_4\} \\ \{\delta_1\} \\ \{\delta_2\} \end{Bmatrix} = \begin{Bmatrix} \{F_4\} \\ \{F_1\} \\ \{F_2\} \end{Bmatrix} \tag{4-19}$$

From Eqs. (4—18) and (4—19), the force $\{F_4\}$ acting on node 4 from element (1) is the same

$$K_{42}\delta_2 + K_{44}\delta_4 + K_{41}\delta_1 = \{F_4\}$$

Writing them into the global stiffness matrix, we can find their positons are exact same.

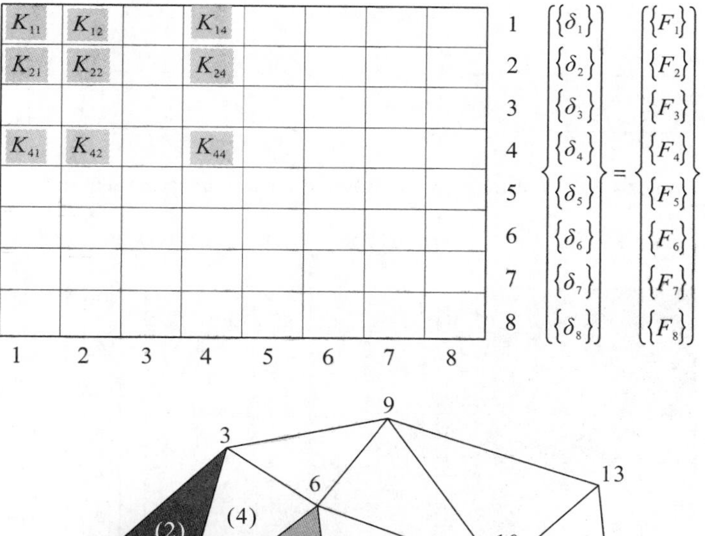

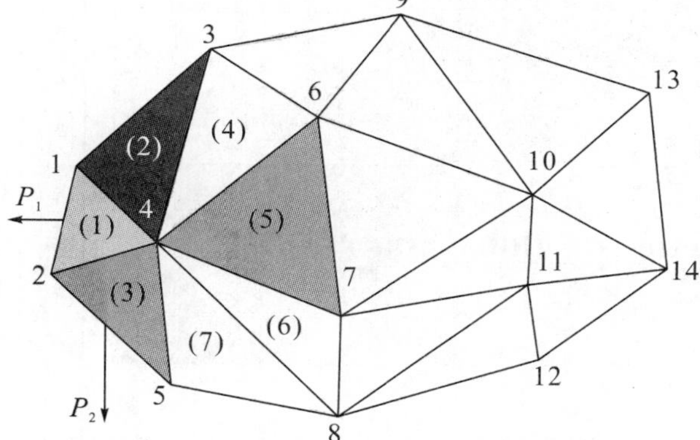

Fig. 4—8 Two loads applied on the domain

(3) For the case that there are two loads applied, the global nodal force vector $\{F\}$ should be

$$\{F\} = \begin{Bmatrix} F_1 \\ F_2 \\ F_3 \\ F_4 \\ F_5 \\ F_6 \\ \vdots \\ F_{13} \\ F_{14} \end{Bmatrix} = \begin{Bmatrix} \{P_1\} \\ \{P_1\}+\{P_2\} \\ 0 \\ 0 \\ \{P_2\} \\ 0 \\ \vdots \\ 0 \\ 0 \end{Bmatrix} \tag{4-20}$$

It should be noted that if a node is free without external load, then the corresponding nodal load is zero in the global nodal force vector. This indicates if no body force is considered, for the case of only one concentrated load, most of the nodal forces equal zero.

(4) For a same structure, different nodal numbering sequences will result in different global stiffness matrix. For the structure shown in Fig. 4−9, there are two different nodal numbering sequences.

Fig. 4−9 A structure with different nodal numbering sequences

For the first case in Fig. 4−9, the global stiffness matrix is

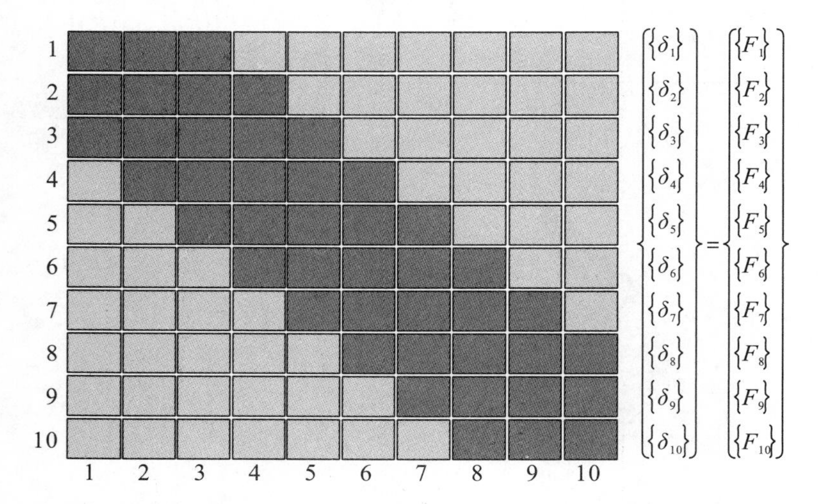

For the second case in Fig. 4−9, the global stiffness matrix is

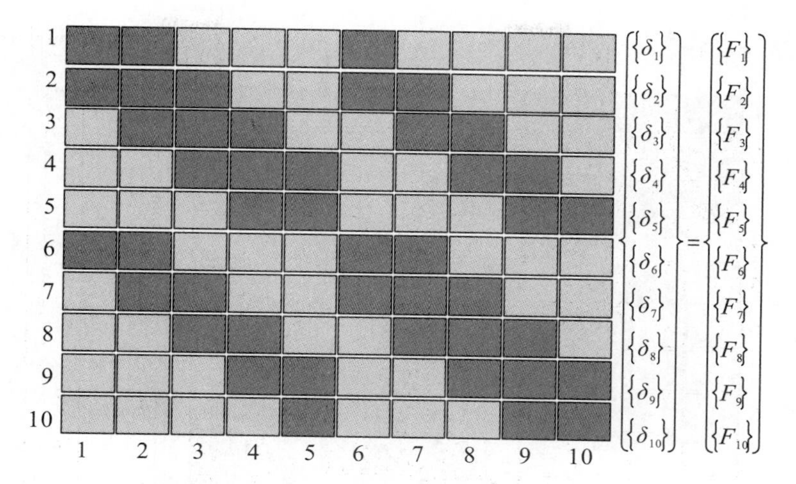

It can be seen that different numbering sequences resulting in different global matrix, and for the first case, all the nonzero terms are clustered near the diagonal. Since we will store the nonzero terms by using half bandwidth method, the first case of numbering method can save the storage largely. Therefore, when we make meshing, the difference between adjacent numbers should be as small as possible.

(5) The global matrix of coefficients is sparse; that is, most of the coefficient are zero. In addition, the nonzero terms are clustered in a narrow band along the main diagonal. This bandedness is characteristic of finite element system equations; it can results in a tremendous saving in both computer time and storage, and means a reduction in cost. Because of these savings, it becomes practical to solve very complicated problems whose system equations contain tens and even hundreds of thousands of unknowns.

(6) In order to describe the properties of bandedness, we use half bandwidth D which includes the term located at the diagonal. For a node (represents the corresponding row number), D can be calculated by the adjacent maximum node number minus this node number and plus 1, and then times 2, i. e.

$$D=(\text{adjacent max node No.} - \text{current No.} + 1)\times \text{node freedom} \qquad (4-21)$$

For 2-D case, D can be written as

$$D=(\text{adjacent max node No.} - \text{current No.} + 1)\times 2 \qquad (4-22)$$

For 3-D case, D can be expressed as

$$D=(\text{adjacent max node No.} - \text{current No.} + 1)\times 3 \qquad (4-23)$$

For a structure shown in Fig. 4-10, we can practice to fill out the global matrix, and to calculate the half bandwidth.

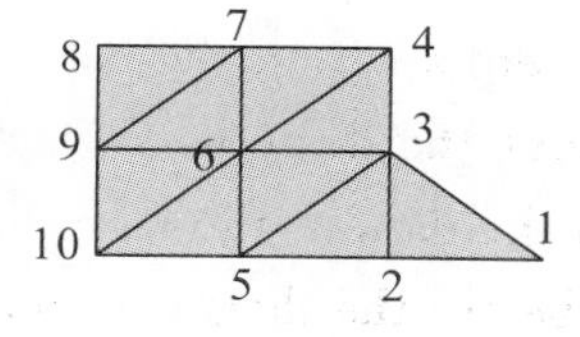

Fig. 4-10 A structure with numbering

The global matrix according to the numbering in Fig. 4—10 can be written as

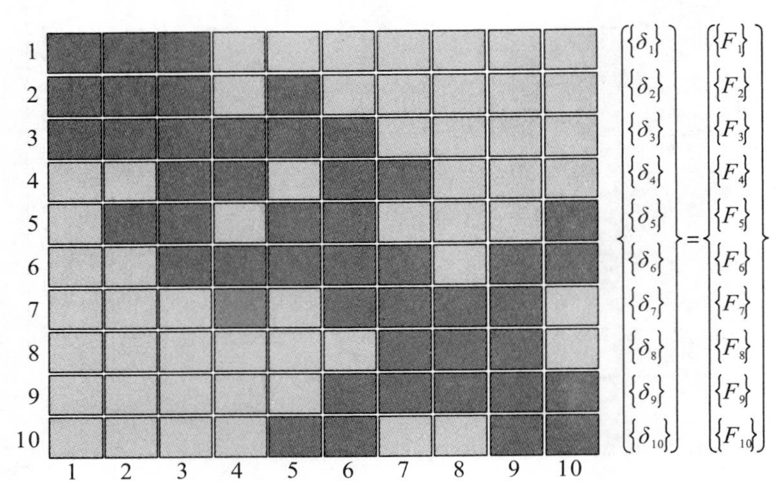

The bandwidth for the second row in Fig. 4—10 can be calculated by D=(5−2+1)× 2=8. The students can find the bandwidth in other rows as an assignment.

(7) The global stiffness matrix has the properties of symmetry and singularity. The global matrix is symmetrical because the element stiffness is symmetrical.

$$[K]^e = \begin{bmatrix} K_{ii} & K_{ij} & K_{ik} \\ K_{ji} & K_{jj} & K_{jk} \\ K_{ki} & K_{kj} & K_{kk} \end{bmatrix}$$

As we install the terms K_{ij} and K_{ji} in the global matrix, they will be installed in the symmetrical position (about the diagonal) again, thus the global matrix must be symmetrical.

4.4 Subroutine of global stiffness matrix

In the previous study, we have used three elements to demonstrate the relationship between the element stiffness matrices and the global stiffness matrix. We find that the global stiffness matrix consists of the three element stiffness matrices. All the terms in an element stiffness matrix should be installed into the global stiffness matrix. Therefore, when an element stiffness matrix is calculated, it should be installed into the global matrix according to the real number of the three nodes.

It is better to only store the terms within the half bandwidth, as shown in Fig. 4—11, because of the symmetry about the diagonal, which can save a large space. In the calculation, we can easily obtain the data in the term below the diagonal. For example, the term K_{41}, which is located below the diagonal and will not be stored in the global matrix, we can use the data of K_{14} instead if need.

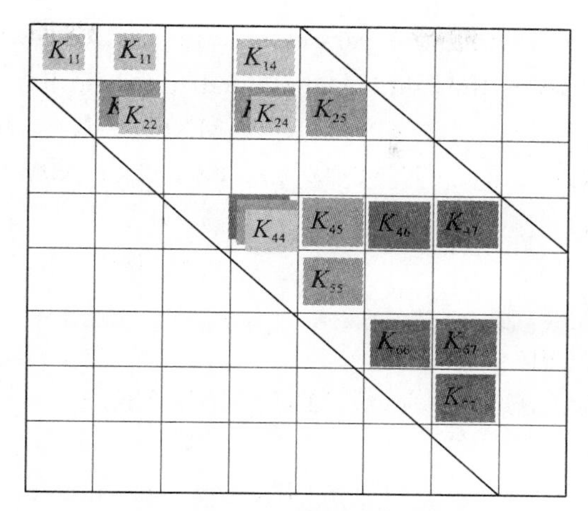

Fig. 4—11 Half bandwidth

Because for each row of an array, the number must start from 1, thus we have to renumber them. For example, the K_{44} has to be stored in K_{41}, and K_{45} has to be stored in K_{42}. Similarly, for other rows, we take the same procedure to renumber them, as shown in Fig. 4—12. Finally the global matrix for an eight-node structure should be the form as shown in Fig. 4—13.

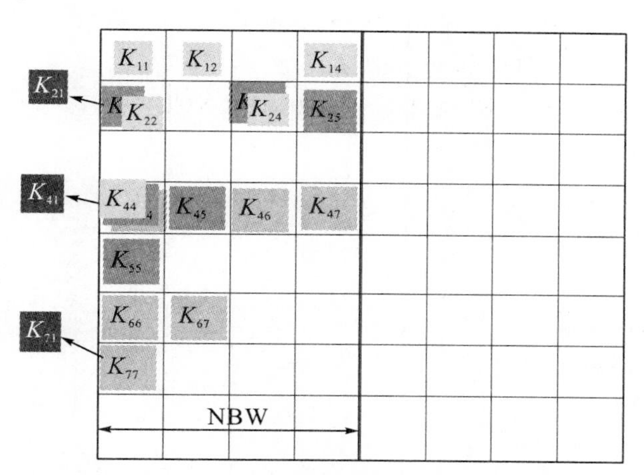

Fig. 4—12 The global stiffness matrix stored

$$
\begin{bmatrix}
k_{11} & k_{12} & \cdots & k_{1r} & \cdots & k_{1NBW} \\
k_{21} & k_{22} & \cdots & k_{2r} & \cdots & k_{2NBW} \\
\vdots & \vdots & & \vdots & & \vdots \\
k_{r1} & k_{r2} & \cdots & k_{rr} & \cdots & k_{rNBW} \\
\vdots & \vdots & & \vdots & & \vdots \\
k_{161} & k_{162} & \cdots & k_{16r} & \cdots & k_{16NBW}
\end{bmatrix}
$$

Fig. 4—13 The form of global stiffness matrix

It can be seen that in final global matrix, the row number is double of the node number, and the column number is NBW.

Next, we will design the global stiffness matrix. The subroutine is named GSM, followed by a bracket containing the parameters transmitted, where GK is the global stiffness matrix, and NBW is the number of the half bandwidth.

```
      Subroutine GSM(NNE, CN, E0, U0, t, MNE, MNN, NN2, NBW, GK)
      Dimension NNE(MNE, 3), CN(MNN, 2), GK(NN2, NBW), S(3, 6), EK(6, 6)
      Do 100 I = 1,NN2          } all terms in GK = 0 because
      Do 100 J = 1,NBW          } some terms in GK = 0
100   GK(I,J) = 0.0             }
      Do 400 NE = 1,MNE
      Call ESM(NE,NNE,CN,E0,U0,t,MNE,MNN,EK,S)
      Do 300 IR = 1,3                       }
      Do 300 I2 = 1,2                       } The row No. in element
      NRE = (IR-1)*2+I2                     } and in global matrix
      NRG = 2*(NNE(NE,IR)-1)+I2             }
      Do 200 JC = 1,3                       }
      Do 200 J2 = 1,2                       } The column No. in
      NCE = 2*(JC-1)+J2                     } element and in global
      NCG = 2*(NNE(NE,JC)-1)+J2             } matrix
      ND = NCG+1-NRG            } below the diagonal
      If (ND. LE. 0) Go to 200  } is not stored
      GK(NRG,ND) = GK(NRG,ND)+EK(NRE,NCE)
200   Continue
300   Continue
400   Continue
      End
```

NNE=Node Number of Element
MNE=Maximum Number of Elements
CN=Coordinates of Nodes
MNN=Maximum Number of Nodes
P=load vector
NN2=2 * MNN
NBW=Number of Bandwidth/2
E0=E; U0=v; t=thickness; Gam=r
GK=Global stiffness matrix
EK=element stiffness matrix
B=strain matrix
S=stress matrix
NRE=No. of row in element matrix
NRG=No. of row in global matrix
NCE=No. of column in element matrix
NCG=No. of column in global matrix
ND=renumbering; e. g. the term K_{44} in the diagonal changes to K_{41}

The students should fully grasp the procedure of installing the element stiffness matrix into the global stiffness matrix.

Assignments:

1. A cantilever beam is divided into four triangular elements, as shown in the following figure, and is subjected to a uniform shear force (the resultant is P) at the right end; the Poisson's ratio μ is 0.25, and the thickness is t. Show the global stiffness matrix.

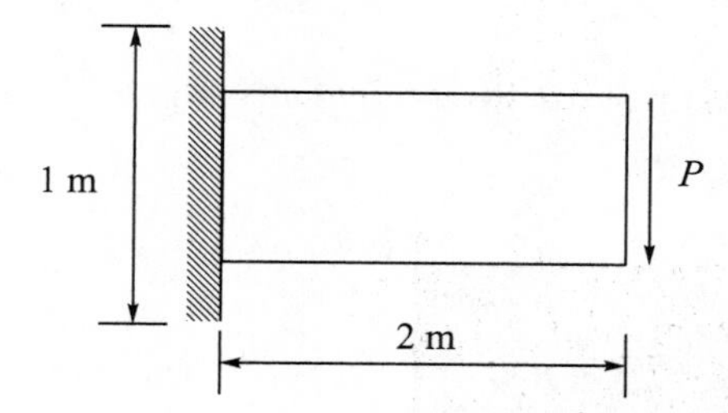

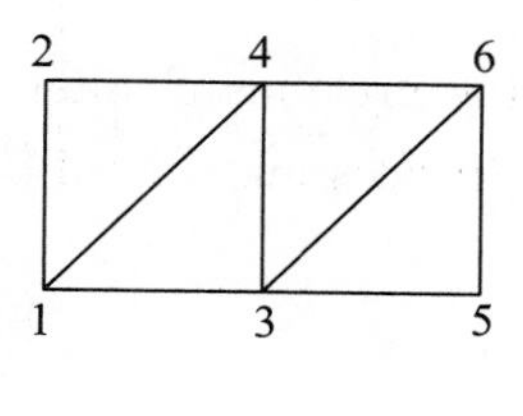

2. A rectangle plate is meshed as shown in the following figure. Please rationally number the nodes and show the nonzero terms in global stiffness matrix.

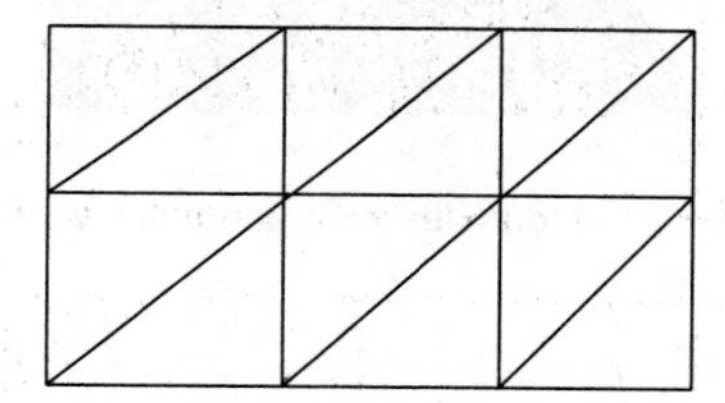

Chapter 5 Boundary conditions and solution of equilibrium equations

From the previous chapter, for a domain with three elements shown in Fig. 5－1, the global stiffness matrix can be written as

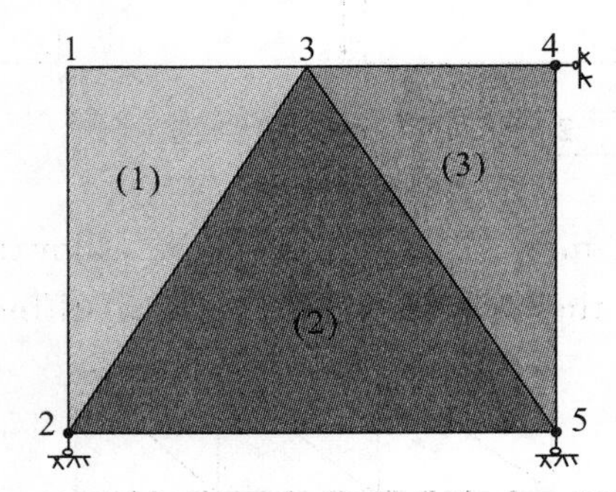

Fig. 5－1 A domain with boundary conditions

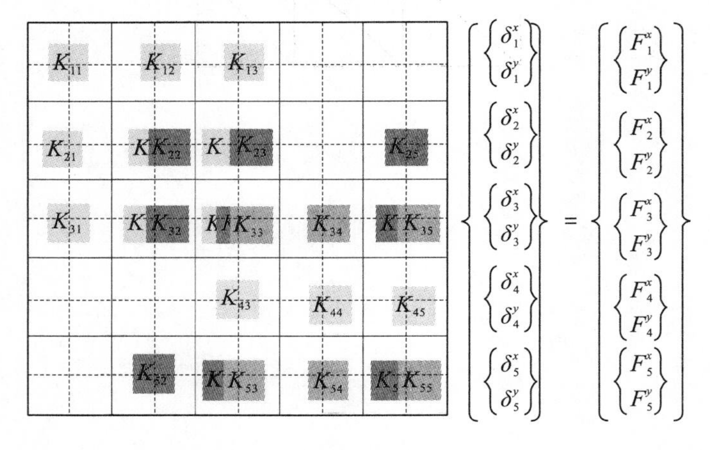

However, we have not considered the boundary conditions yet. In Fig. 5－1, at node 2, the displacements in x and y directions are fixed; at node 4, the displacement in x direction is fixed; and at node 5, the displacement in y direction is fixed. Under such scenario, we have to make sure that in the displacement solution, $\delta_2^x = 0$, $\delta_2^y = 0$, $\delta_4^x = 0$ and $\delta_5^y = 0$, as shown in Eq. (5－1).

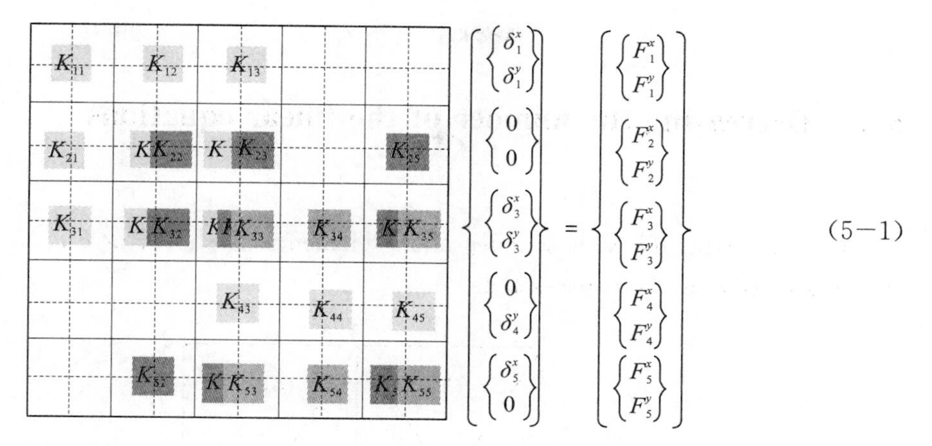

(5−1)

Therefore, before using solver to obtain the solution, we have to adjust the global equations according to the displacement conditions since some displacement components are known already. There are three methods which can be easily realized in computer programming to adjust the global equations.

1. Multiplying a large number;
2. Decreasing the number of the linear equations;
3. Changing the diagonal term to one.

5.1 Multiplying a large number

Suppose the displacement δ_r (in the rth row) is known, i. e. $\delta_r = c_r$ (where c_r is a constant). We multiply a very large number, say $10^8 \sim 10^{10}$, then the term in the corresponding diagonal can be rewritten as $Nk_{rr}c_r$.

$$\begin{bmatrix} k_{11} & k_{12} & \cdots & k_{1r} & \cdots & k_{116} \\ k_{21} & k_{22} & \cdots & k_{2r} & \cdots & k_{216} \\ \vdots & \vdots & & \vdots & & \vdots \\ k_{r1} & k_{r2} & \cdots & Nk_{rr} & \cdots & k_{r16} \\ \vdots & \vdots & & \vdots & & \vdots \\ k_{161} & k_{162} & \cdots & k_{16r} & \cdots & k_{1616} \end{bmatrix} \begin{Bmatrix} \delta_1 \\ \delta_2 \\ \vdots \\ \delta_r \\ \vdots \\ \delta_6 \end{Bmatrix} = \begin{Bmatrix} F_1 \\ F_2 \\ \vdots \\ Nk_{rr}c_r \\ \vdots \\ F_{16} \end{Bmatrix} \tag{5-2}$$

The load F_r in the row in Eq. (5−2) changes to $Nk_{rr}c_r$, then for the rth row

$$k_{r1}\delta_1 + k_{r2}\delta_2 + \cdots + Nk_{rr}\delta_r + \cdots + k_{r16}\delta_{16} = Nk_{rr}c_r \tag{5-3}$$

Because N is very large, the other terms in Eq. (5−3) can be ignored, and then we have

$$Nk_{rr}\delta_r = Nk_{rr}c_r$$

Under such operation, we can guarantee that $\delta_r = c_r$, and meanwhile, the other unknowns are not affected by this operations.

5.2 Decreasing the number of the linear equations

For the structure with 8 nodes and 7 elements as shown in Fig. 5−2, the global stiffness matrix can be expressed as

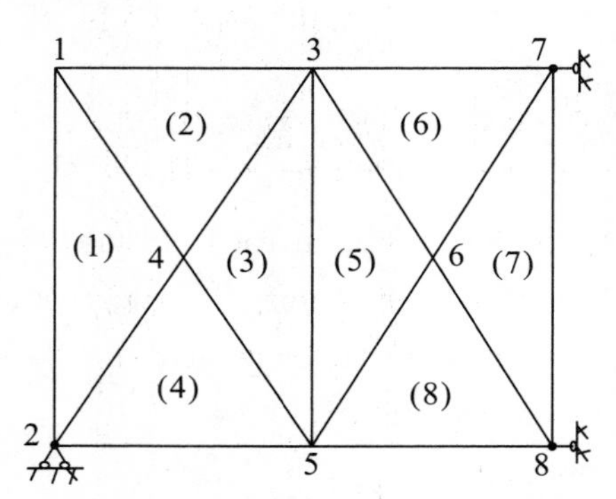

Fig. 5−2 A domain with four displacements known

$$\begin{bmatrix} k_{11} & k_{12} & \cdots & k_{113} & \cdots & k_{116} \\ k_{21} & k_{22} & \cdots & k_{213} & \cdots & k_{216} \\ \vdots & \vdots & & \vdots & & \vdots \\ k_{r1} & k_{r2} & \cdots & k_{r13} & \cdots & k_{r16} \\ \vdots & \vdots & & \vdots & & \vdots \\ k_{131} & k_{132} & \cdots & k_{1313} & \cdots & k_{1316} \\ \vdots & \vdots & & \vdots & & \vdots \\ k_{161} & k_{162} & \cdots & k_{163} & \cdots & k_{1616} \end{bmatrix} \begin{Bmatrix} \delta_1 \\ \delta_2 \\ \vdots \\ \delta_r \\ \vdots \\ \delta_{13} \\ \vdots \\ \delta_{16} \end{Bmatrix} = \begin{Bmatrix} F_1 \\ F_2 \\ \vdots \\ F_r \\ \vdots \\ F_{13} \\ \vdots \\ F_{16} \end{Bmatrix} \tag{5−4}$$

The displacements $\delta_3, \delta_4, \delta_{13}$ and δ_{15} in the 3th, 4th, 13th and 15th rows are known, and we regroup them together as shown in Eq. (5−5).

$$\left[\begin{array}{ccccc|cccc} k_{11} & k_{12} & k_{15} & \cdots & k_{116} & k_{13} & k_{14} & k_{113} & k_{115} \\ k_{21} & k_{22} & k_{25} & \cdots & k_{216} & k_{23} & k_{24} & k_{213} & k_{215} \\ \vdots & \vdots & \vdots & [K_{aa}] & \vdots & \vdots & & \vdots[K_{ab}]\vdots & \vdots \\ k_{161} & k_{162} & k_{165} & \cdots & k_{1616} & k_{163} & k_{164} & k_{1613} & k_{1615} \\ \hline k_{31} & k_{32} & k_{35} & \cdots & k_{316} & k_{33} & k_{34} & k_{313} & k_{315} \\ k_{41} & k_{42} & k_{45} & \cdots & k_{416} & k_{43} & k_{44} & k_{413} & k_{415} \\ & & & [K_{ba}] & & & & [K_{bb}] & \\ k_{131} & k_{132} & k_{135} & \cdots & k_{1316} & k_{133} & k_{134} & k_{1313} & k_{1315} \\ k_{151} & k_{152} & k_{155} & \cdots & k_{1516} & k_{153} & k_{154} & k_{1513} & k_{1515} \end{array}\right] \begin{Bmatrix} \delta_1 \\ \delta_2 \\ \delta_5 \\ \vdots \\ \delta_{16} \\ \hline \delta_3 \\ \delta_4 \\ \delta_{13} \\ \delta_{15} \end{Bmatrix} = \begin{Bmatrix} F_1 \\ F_2 \\ F_5 \\ \vdots \\ F_{16} \\ \hline F_3 \\ F_4 \\ F_{13} \\ F_{15} \end{Bmatrix} \tag{5−5}$$

Eq. (5−5) can be simplified as

$$\begin{bmatrix} [K_{aa}] & [K_{ab}] \\ [K_{ba}] & [K_{bb}] \end{bmatrix} \begin{Bmatrix} \{\delta_a\} \\ \{\delta_b\} \end{Bmatrix} = \begin{Bmatrix} \{F_a\} \\ \{F_b\} \end{Bmatrix} \tag{5−6}$$

where $\{\delta_b\}$ is the known displacements. From Eq. (5−6), the number of the linear equations can be decreased as

$$[K_{aa}]\{\delta_a\} = \{F_a\} - [K_{ab}]\{\delta_b\} \tag{5-7}$$

From Eq. (5−5), we can find that the original 16 linear equations are decreased to 12 linear equations.

5.3 Changing the diagonal term to one

Suppose the displacement δ_r (in the rth row) is known, i. e. $\delta_r = c_r$ (where c_r is a constant). We can change the diagonal term in the rth row to 1, and the other terms are changed as shown in Eq. (5−8).

$$\begin{bmatrix} k_{11} & k_{12} & \cdots & 0 & \cdots & k_{116} \\ k_{21} & k_{22} & \cdots & 0 & \cdots & k_{216} \\ \vdots & \vdots & & \vdots & & \vdots \\ 0 & 0 & \cdots & 1 & \cdots & 0 \\ \vdots & \vdots & & \vdots & & \vdots \\ k_{161} & k_{162} & \cdots & 0 & \cdots & k_{1616} \end{bmatrix} \begin{Bmatrix} \delta_1 \\ \delta_2 \\ \vdots \\ \delta_r \\ \vdots \\ \delta_{16} \end{Bmatrix} = \begin{Bmatrix} F_1 - k_{1r}c_r \\ F_2 - k_{2r}c_r \\ \vdots \\ c_r \\ \vdots \\ F_{16} - k_{16r}c_r \end{Bmatrix} \tag{5-8}$$

For the first row, the linear equation can be written as

$$k_{11}\delta_1 + k_{12}\delta_2 + \cdots + Nk_{116}\delta_{16} = F_1 - k_{1r}c_r \tag{5-9}$$

It can be seen that if we move the term $k_{1r}c_r$ to the left side, the equation is the same as before the adjustment. Similarly, for other rows, we can get the same results.

If $\delta_r = c_r = 0$, then Eq. (5−8) can be rewritten as

$$\begin{bmatrix} k_{11} & k_{12} & \cdots & 0 & \cdots & k_{116} \\ k_{21} & k_{22} & \cdots & 0 & \cdots & k_{216} \\ \vdots & \vdots & & \vdots & & \vdots \\ 0 & 0 & \cdots & 1 & \cdots & 0 \\ \vdots & \vdots & & \vdots & & \vdots \\ k_{161} & k_{162} & \cdots & 0 & \cdots & k_{1616} \end{bmatrix} \begin{Bmatrix} \delta_1 \\ \delta_2 \\ \vdots \\ \delta_r \\ \vdots \\ \delta_{16} \end{Bmatrix} = \begin{Bmatrix} F_1 \\ F_2 \\ \vdots \\ 0 \\ \vdots \\ F_{16} \end{Bmatrix} \tag{5-10}$$

The change from the original global matrix to the final one is shown in Fig. 5−3.

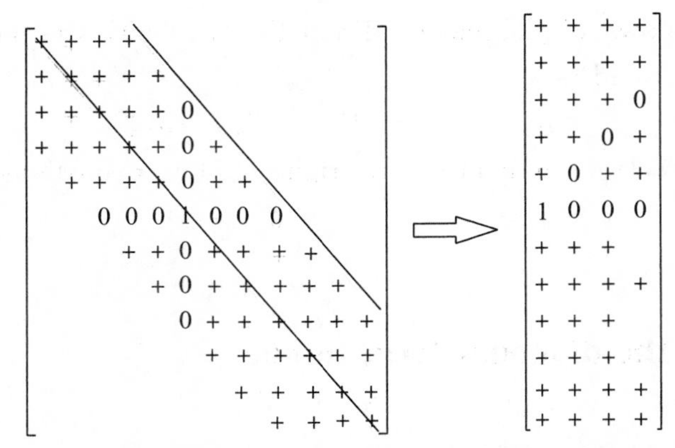

Fig. 5−3 The change from the original global matrix to the final one

From Fig. 5−3, one can find when adjust the global matrix using the method of changing diagonal term to one, we should change the terms in the inclined column to zero, not the vertical column.

5.4 Subroutine of adjusting global matrix

Based on the method of changing diagonal term to 1, the subroutine is designed as follows. The subroutine is named AGM, followed by the parameters transmitted in a bracket. NDK is the number of displacement known, and the array NR is the number of rows which need to adjust.

```
      Subroutine AGM(GK, NBW, NR, NDK, NN2, P)
      Dimension GK(NN2, NBW), NR(NDK), P(NN2)
      Do 600 I=1, NDK   ➡   [the number of displacement known]
      M=NR(I)   ➡   [the number of rows]
      GK(M, 1)=1.0   ➡   [the term in the diagonal=1]
      Do 100 J=2, NBW
100   GK(M, j)=0.0   ➡   [the terms in the same row=0]
      Ju=Min(M, NBW)   ➡   [select the minimum between the row number and NBW]
      Do 200 J=2, Ju
200   GK(M-J+1, J)=0.0   ➡   [the inclined terms=0]
      P(M)=0.0
600   Continue
      Return
      End
```

P=global load vector
NN2=2 * MNN
NBW=Number of Bandwidth/2
GK=Global stiffness Matrix
NDK=Number of Displacement Known
NR=Number of Row

Discussion: in the above subroutine, we have selected the small number between the M (row number) and the NBW (half bandwidth), i. e. "Ju=Min(M, NBW)". This operation can be explained as follows.

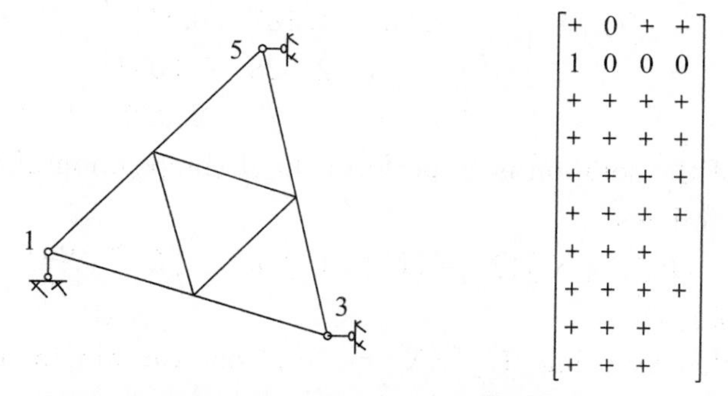

Fig. 5-4 A structure and the corresponding adjustment for node 1

For the node 1 shown in Fig. 5-4, the second row in the global matrix needs to adjust. The NBW in the global is 4, but only one term k_{12} in the inclined column needs to change to zero. In the subroutine, the variable M is 2, thus between M and NBW, M is selected in the circulation.

5.5 Solver

One of the germane steps in a FE analysis is the solution of equilibrium equations. The efficiency of the analysis depends largely on the numerical procedure used. The classical methods existing are: direct solution methods and indirect methods.

Direct solution methods are much more effective in many applications, and the most popular direct solution method is Gauss elimination. In addition, Cholesky decomposition method is another efficient method, and it is first introduced here.

5.5.1 Cholesky decomposition method

Cholesky decomposition method requires the matrix must be symmetric, positive definite. For such a matrix, one can have

$$[K]=[U]^{\mathrm{T}}[U] \tag{5-11}$$

where is $[K]$ symmetric, positive definite matrix, and $[U]$ is upper triangular matrix and it can easily determined. For example,

$$[K]=\begin{bmatrix}5 & 2 & 1 & 1\\ 2 & 10 & 4 & 3\\ 1 & 4 & 8 & 2\\ 1 & 3 & 2 & 12\end{bmatrix}=\begin{bmatrix}u_{11} & & & \\ u_{12} & u_{22} & & \\ u_{13} & u_{23} & u_{33} & \\ u_{14} & u_{24} & u_{34} & u_{44}\end{bmatrix}\begin{bmatrix}u_{11} & u_{12} & u_{13} & u_{14}\\ & u_{22} & u_{23} & u_{24}\\ & & u_{33} & u_{34}\\ & & & u_{44}\end{bmatrix} \tag{5-12}$$

We can have

$$[U]=\begin{bmatrix}\sqrt{5} & \frac{1}{\sqrt{5}} & \frac{2}{\sqrt{5}} & \frac{1}{\sqrt{5}}\\ & 3.033 & 1.187 & 0.857\\ & & 2.528 & 0.3096\\ & & & 3.312\end{bmatrix} \tag{5-13}$$

The first step of the solution is to perform the decomposition of the above equation. The procedure is as follows

$$\{F\}=[K]\{\delta\}=[U]^{\mathrm{T}}[U]\{\delta\}=[U]^{\mathrm{T}}\{X\} \tag{5-14}$$

where $\{X\}=[U]\{\delta\}$.

It can be seen that by using $[U]^{\mathrm{T}}\{X\}=\{F\}$, one can obtain the matrix $\{X\}$, and then use the equation $\{X\}=[U]\{\delta\}$, we can obtain the solution of $\{\delta\}$ by back substituting method.

For example, the global matrix is

$$\begin{bmatrix}5 & 2 & 1 & 1\\ 2 & 10 & 4 & 3\\ 1 & 4 & 8 & 2\\ 1 & 3 & 2 & 12\end{bmatrix}\begin{bmatrix}\delta_1\\ \delta_2\\ \delta_3\\ \delta_4\end{bmatrix}=\begin{bmatrix}3\\ 2\\ 6\\ 4\end{bmatrix} \tag{5-15}$$

According to $[K]=[U]^{\mathrm{T}}[U]$, one can obtain the upper triangular matrix $[U]$ in Eq. (5−13). From $[U]^{\mathrm{T}}\{X\}=\{F\}$, we can easily get the solution $\{X\}$ as

$$\begin{bmatrix}\sqrt{5} & & & \\ \frac{2}{\sqrt{5}} & 3.033 & & \\ \frac{1}{\sqrt{5}} & 1.187 & 2.528 & \\ \frac{1}{\sqrt{5}} & 0.857 & 0.3096 & 3.312\end{bmatrix}\begin{Bmatrix}x_1\\ x_2\\ x_3\\ x_4\end{Bmatrix}=\begin{Bmatrix}3\\ 2\\ 6\\ 4\end{Bmatrix}\longrightarrow\begin{Bmatrix}x_1\\ x_2\\ x_3\\ x_4\end{Bmatrix}=\begin{Bmatrix}\frac{3}{\sqrt{5}}\\ 0.264\\ 2.012\\ 0.77\end{Bmatrix} \tag{5-16}$$

Based on $\{X\}=[U]\{\delta\}$, we have

$$\begin{Bmatrix}x_1\\ x_2\\ x_3\\ x_4\end{Bmatrix}=\begin{Bmatrix}\frac{3}{\sqrt{5}}\\ 0.264\\ 2.012\\ 0.77\end{Bmatrix}=[U]\{\delta\}=\begin{bmatrix}\sqrt{5} & \frac{1}{\sqrt{5}} & \frac{2}{\sqrt{5}} & \frac{1}{\sqrt{5}}\\ & 3.033 & 1.187 & 0.857\\ & & 2.528 & 0.3096\\ & & & 3.312\end{bmatrix}\begin{Bmatrix}\delta_1\\ \delta_2\\ \delta_3\\ \delta_4\end{Bmatrix} \tag{5-17}$$

Finally, we have

$$\begin{Bmatrix} \delta_1 \\ \delta_2 \\ \delta_3 \\ \delta_4 \end{Bmatrix} = \begin{Bmatrix} 0.5115 \\ -0.2787 \\ 0.7674 \\ 0.2325 \end{Bmatrix} \tag{5-18}$$

5.5.2 Gauss elimination method

This is the most effective direct solution technique and can be applied to almost any set of linear simultaneous equations. The mathematical procedure is simple and stepwise. The approach is to reduce the stiffness matrix into an upper triangular matrix by performing a series of row operations. This enables the unknown coefficients to be calculated easily in succession by backsubstituting.

Example: solve the following equilibrium equations for the unknowns; in the following, the number in a bracket represents the row number.

$$\begin{bmatrix} 4 & -6 & 2 & 0 \\ -6 & 24 & 0 & 6 \\ 2 & 0 & 8 & 2 \\ 0 & 6 & 2 & 4 \end{bmatrix} \begin{bmatrix} \delta_1 \\ \delta_2 \\ \delta_3 \\ \delta_4 \end{bmatrix} = \begin{bmatrix} 0 \\ 6 \\ 0 \\ 0 \end{bmatrix} \tag{5-19}$$

Step 1: (1) $\times \frac{6}{4}$ + (2); (1) $\times (-\frac{1}{2})$ + (3); This will result in zeros in the first column except for the first row.

$$\begin{bmatrix} 4 & -6 & 2 & 0 \\ 0 & 15 & 3 & 6 \\ 0 & 3 & 7 & 2 \\ 0 & 6 & 2 & 4 \end{bmatrix} \begin{bmatrix} \delta_1 \\ \delta_2 \\ \delta_3 \\ \delta_4 \end{bmatrix} = \begin{bmatrix} 0 \\ 6 \\ 0 \\ 0 \end{bmatrix} \tag{5-20}$$

Step 2: (2) $\times (-\frac{3}{15})$ + (3); (2) $\times (-\frac{6}{15})$ + (4), and the results is

$$\begin{bmatrix} 4 & -6 & 2 & 0 \\ 0 & 15 & 3 & 6 \\ 0 & 0 & \frac{32}{5} & \frac{4}{5} \\ 0 & 0 & \frac{4}{5} & \frac{8}{5} \end{bmatrix} \begin{bmatrix} \delta_1 \\ \delta_2 \\ \delta_3 \\ \delta_4 \end{bmatrix} = \begin{bmatrix} 0 \\ 6 \\ -\frac{6}{5} \\ -\frac{12}{5} \end{bmatrix} \tag{5-21}$$

Step 3: (3) $\times (-\frac{1}{8})$ + (4), and the results is

$$\begin{bmatrix} 4 & -6 & 2 & 0 \\ 0 & 15 & 3 & 6 \\ 0 & 0 & \frac{32}{5} & \frac{4}{5} \\ 0 & 0 & 0 & \frac{3}{2} \end{bmatrix} \begin{bmatrix} \delta_1 \\ \delta_2 \\ \delta_3 \\ \delta_4 \end{bmatrix} = \begin{bmatrix} 0 \\ 6 \\ -\frac{6}{5} \\ -\frac{9}{4} \end{bmatrix} \tag{5-22}$$

It can be seen that the above matrix is an upper triangular matrix, and we can now get the unknowns by backsubstituting method starting from the last one, we have

$$\delta_4=\frac{-\frac{9}{4}}{\frac{3}{2}}=-\frac{3}{2}\qquad \delta_3=\frac{-\frac{6}{5}-\delta_4\cdot\frac{4}{5}}{\frac{32}{5}}=0$$

$$\delta_2=\frac{6-3\delta_3-6\delta_4}{15}=1\qquad \delta_1=\frac{6\delta_2-2\delta_3}{4}=\frac{3}{2}$$

5.5.3 Subroutine of solver

Based on the half bandwidth global matrix, the solver subroutine is design as

```
      Subroutine Solver(GK, P, NBW, MNN, NN2)
      Dimension GK(NN2, NBW), P(NN2)
      Do 50 K=1, NN2-1
      if((NN2-K-NBW+1).GT.0.0) go to 20
      Im=NN2
      Go to 30
20    Im=K+NBW-1
30    K1=K+1
      Do 50 I=K1, Im
      L=I-K+1
      C=GK(K, L)/GK(K, 1)
      LD1=NBW-L+1
      DO 40 J=1, LD1
      M=J+I-K
40    GK(I, J)=GK(I, J)-C*GK(K, M)
      P(I)=P(I)-C*P(K)
50    CONTINUE
      P(NN2)=P(NN2)/GK(NN2, 1)
      DO 100 I=NN2-1, 1, -1
      If((NBW-NN2+I -1).GT.0)Goto 70
      JU=NBW
      Go to 80
70    JU=NN2-I+1
80    DO 90 J=2, JU
      LH=J+I-1
90    P(I)=P(I)-GK(I, J)*P(LH)
      P(I)=P(I)/GK(I, 1)
100   CONTINUE
```

GK=Global stiffness Matrix
P=global load vector
NBW=Number of Bandwidth/2
MNN=Maximum Number of Nodes
NN2=2 * MNN

```
      WRITE(9, 500)(I, P(2*I-1), P(2*I), I=1, MNN)
500   format(2x,'Node no.  =',5I, 3x,'D-x=',f20.5,
     ! 3x,'D-y=',f20.5, /)
      END
```

It should be noted that in order to save the space, the solutions of the displacements are stored in the P array which originally were used to store the loads.

Assignments:

1. How stress boundary conditions and displacement conditions are considered in finite element method? How many methods can be used to adjust the global stiffness matrix?

2. An equilibrium equation is shown in the following. Solve the displacements by using Gauss elimination method.

$$\begin{bmatrix} 2 & 4 & 2 & 0 \\ 4 & 4 & 0 & 1 \\ 2 & 0 & 8 & -2 \\ 0 & 1 & -2 & 4 \end{bmatrix} \begin{bmatrix} \delta_1 \\ \delta_2 \\ \delta_3 \\ \delta_4 \end{bmatrix} = \begin{bmatrix} 1 \\ 0 \\ 4 \\ 0 \end{bmatrix}$$

3. An equilibrium equation is shown in the following. Solve the displacements by using Cholesky decomposition method.

$$\begin{bmatrix} 2 & -2 & 4 & -4 \\ -2 & 6 & 0 & 2 \\ 4 & 0 & 16 & 1 \\ -4 & 2 & 1 & 40 \end{bmatrix} \begin{bmatrix} \delta_1 \\ \delta_2 \\ \delta_3 \\ \delta_4 \end{bmatrix} = \begin{bmatrix} 2 \\ 0 \\ 1 \\ 0 \end{bmatrix}$$

4. The loads and the boundary conditions of a square plate are shown in the following figure, and the square plate is divided into two triangular elements. The Poisson's ratio is $\mu=0.2$, elastic modulus is $E = 200$ GPa, thickness is $t = 1.0$, $a=1$ m and $P=100$ kN. Calculate the displacement and stresses of each node.

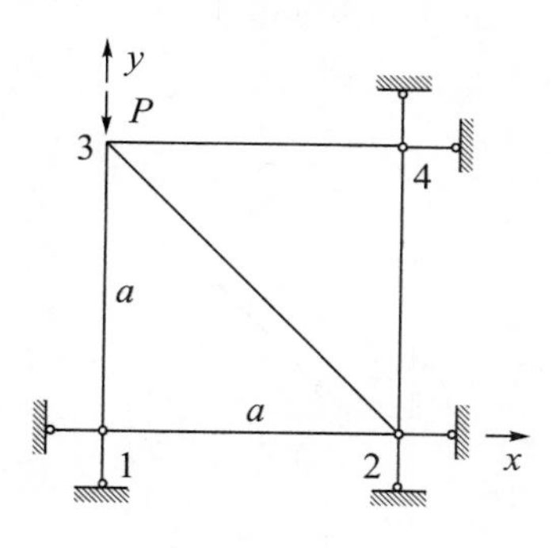

Chapter 6　Subroutine of nodal stresses and main program

6.1　The calculation method of nodal average stresses

By using Gauss elimination method, we can get the solutions of the nodal displacements, and for an eight node structure shown in Fig. 6－1, one can obtain 16 displacements.

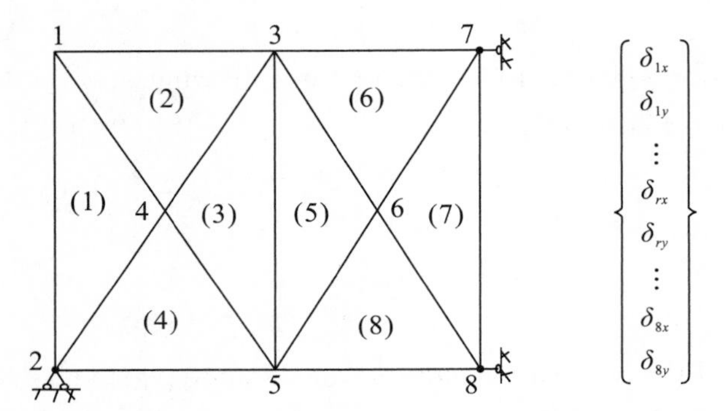

$$\begin{Bmatrix} \delta_{1x} \\ \delta_{1y} \\ \vdots \\ \delta_{rx} \\ \delta_{ry} \\ \vdots \\ \delta_{8x} \\ \delta_{8y} \end{Bmatrix}$$

Fig. 6－1　Eight node structure and the displacements solutions

From Chapter 1, we know the relationship between the element stresses and the three node displacements, and it can be obtained by

$$\{\sigma\} = [D]\{\varepsilon\} = [D][B]\{\delta\}^e = [S]\{\delta\}^e$$

[shown in Eq. (1－17)]

where

$$[D] = \frac{E}{1-\nu^2}\begin{bmatrix} 1 & \nu & 0 \\ \nu & 1 & 0 \\ 0 & 0 & \frac{1-\nu}{2} \end{bmatrix}$$

$$[B] = \frac{1}{2A}\begin{bmatrix} b_i & 0 & b_j & 0 & b_k & 0 \\ 0 & c_i & 0 & c_j & 0 & c_k \\ c_i & b_i & c_j & b_j & c_k & b_k \end{bmatrix}$$ [shown in Eq. (1－13)]

and

$$[S]=\frac{E}{2A(1-\nu^2)}\begin{bmatrix} b_i & \nu c_i & b_j & \nu c_j & b_k & \nu c_k \\ \nu b_i & c_i & \nu b_j & c_j & \nu b_k & c_k \\ \frac{1-\nu}{2}c_i & \frac{1-\nu}{2}b_i & \frac{1-\nu}{2}c_j & \frac{1-\nu}{2}b_j & \frac{1-\nu}{2}c_k & \frac{1-\nu}{2}b_k \end{bmatrix}$$

[shown in Eq. (1−18)]

where

$$\begin{cases} a_i = x_j y_k - x_k y_j \\ b_i = y_j - y_k \qquad (i,j,k) \\ c_i = -x_j + x_k \end{cases}$$

The element stresses can be expressed as

$$\begin{Bmatrix} \sigma_x \\ \sigma_y \\ \tau_{xy} \end{Bmatrix}=\frac{E}{2A(1-\nu^2)}\begin{bmatrix} b_i & \nu c_i & b_j & \nu c_j & b_k & \nu c_k \\ \nu b_i & c_i & \nu b_j & c_j & \nu b_k & c_k \\ \frac{1-\nu}{2}c_i & \frac{1-\nu}{2}b_i & \frac{1-\nu}{2}c_j & \frac{1-\nu}{2}b_j & \frac{1-\nu}{2}c_k & \frac{1-\nu}{2}b_k \end{bmatrix}\begin{Bmatrix} \delta_{ix} \\ \delta_{iy} \\ \delta_{jx} \\ \delta_{jy} \\ \delta_{kx} \\ \delta_{ky} \end{Bmatrix}$$

[shown in Eq. (1−19)]

From Eq. (1−19), we can find that if the six displacements of an element is known, the three stress components of the element can be easily calculated.

There are two issues we need to consider: first is how to collect the six displacements from the global displacements group, as shown in Fig. 6−2. The six displacements should be collected according to the three node number. For example, for element (4) in Fig. 6−1, the i, j and k number are 2, 5 and 4, and the six displacements of this element should be

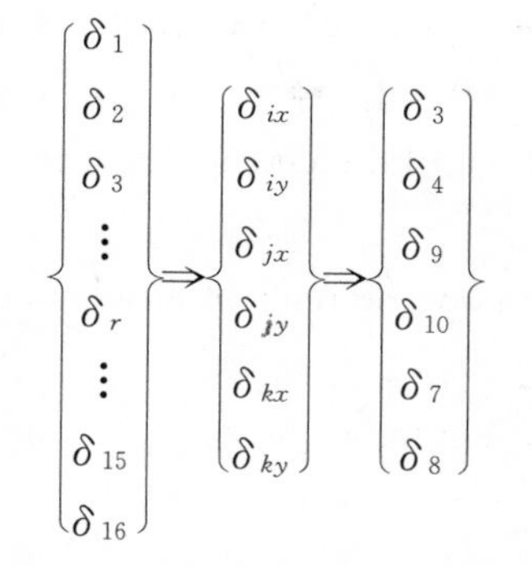

Fig. 6−2 The displacement relation

The second is how to design the program to calculate the nodal stresses because one node may connect with several elements, such as the node 4 in Fig. 6−1. Under such scenario, we have to use the average stresses of the surrounding elements, i. e. the average stresses of elements 1, 2, 3 and 4, to express the node 4 stresses.

6.2 Subroutine of nodal stress

The subroutine is named stress, and the parameters transmitted are in the following bracket. It should be noted that the P array right now is used to store all the nodal displacements. We use DE array to store the six displacements of an element. We use ST array to store nodal stresses (each node has three stress components). We use NT (counter) to store the repeating times of a node because one node may connect with several elements, and the repeating times will be used to calculate the average nodal stresses. We use SE to store the element stresses.

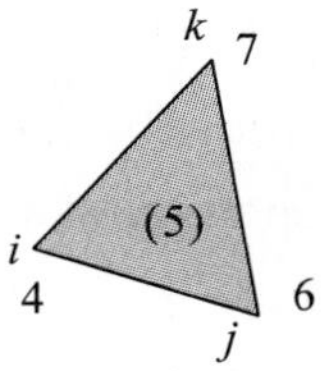

```
     Subroutine Stress(NNE, CN, E0, U0, t, MNE, MNN, NN2, P)
     Dimension NNE(MNE,3),CN(MNN,2),EK(6,6),S(3,6),DE(6),SE(3),
               ST(MNN,3),P(NN2),NT(MNN)
     Do 100 NN = 1,MNN
     NT(NN) = 0  ➡  [initial number of counter = 0]
     Do 100,J = 1,3
100  ST(NN,J) = 0.0  ➡  [initial stresses = 0]
     Do 400 NE = 1,MNE  ➡  [element circulation]
     Call ESM(NE,NNE,CN,E0,U0,t,MNE,MNN,EK,S)
     Do 200 I = 1,3
     Do 200 J = 1,2
     NRE = 2 * (I - 1) + J  ➡  [No. of row in element]
     NRG = 2 * (NNE(NE,I) - 1) + J  ➡  [No. of row in global]
200  DE(NRE) = P(NRG)  ➡  [find element displacement]
     i = NNE(NE,1) ⎫
     j = NNE(NE,2) ⎬  three node number of an element
     k = NNE(NE,3) ⎭
     Do 400 M = 1,3  ➡  [row number]
     SE(M) = 0.0
     IR = NNE(NE,M)     ⎫
                        ⎬  count the times of a node
     NT(IR) = NT(IR) + 1⎭
     Do 300 N = 1,6  ➡  [column number]
300  SE(M) = SE(M) + S(M,N) * DE(N)  ➡  [calculate element stress]
     ST(i,M) = ST(i,M) + SE(M) ⎫
     ST(j,M) = ST(j,M) + SE(M) ⎬store stress into the three nodes
     ST(k,M) = ST(k,M) + SE(M) ⎭
```

k 7
(5)
i
4
6
j

```
400  Continue
     Do 800 I = 1,MNN                    ┐
     Do 800 J = 1,3                      │
     ST(I,J) = ST(I,J) / NRT(I)          ├calculate average nodal stress
800  write(9,900) I,ST(I,J)              ┘
900  format(2x,I6,3x,3F20.5,/)
     Return
     End
```

NNE=Node Number of Elements
CN=Coordinates of Nodes
MNE=Maximum Number of Elements
MNN=Maximum Number of Nodes
NN2=2 * MNN
E0=E; U0=v; t=thickness
EK=element stiffness matrix
S=stress matrix
NRE=No. of row in element
NRG=No. of row in global
ST=nodal stress
SE=element stress
DE=element displacement
NT=No. of times of a node
P=global displacements

6.3 Main program

Main program usually has the following functions:

1. Make definition of the arrays and parameters used in the program;
2. Input the initial data which will be used in the subsequent calculation;
3. Call the subroutines to perform calculations;
4. Output the calculation results.

We design the main program for finite element code as follows:

Main program

```
Dimension NNE (MNE,3),CN (MNN,2),P(NN2),GK(NN2,NBW),NR(x)
open (5,file = 'Input.dat')
open (9,file = 'result.out')
read (5, * ) MNN,MNE,NL,NBW,NDK
read (5, * ) E0,U0,t,Gam
```

```
      read (5, * ) ((NNE(i,j),j = 1,3),i = 1,MNE)
      read (5, * ) ((CN (i,j),j = 1,2),i = 1,MNN)
      read(5, * ) (NR(I),I = 1,NDK)
      NN2 = NN * 2
      Do 100 I = 1,NN2
100   P(I) = 0.0
      If (NL. Eq. 0) go to 300
      Do 200 I = 1,NL
200   read(5, * ) NRL,P(NRL)
300   Call GSM(NNE,CN,E0,U0,t,MNE,MNN,NN2,NBW,GK)
      Call Load ((NNE,MNE,CN,MNN,NN2,P,Gam)
      Call AGM(GK,NBW,NR,NDK,NN2,P)
      Call Solver (GK,P,NBW,MNN,NN2)
      Call Stress (NNE,CN,E0,U0,t,MNE,MNN,NN2,P)
      Stop
      End
```

NNE=Node Number of Elements
CN=Coordinates of Nodes
MNE=Maximum Number of Elements
MNN=Maximum Number of Nodes
NN2=2 * MNN
NBW=Number of Bandwidth/2
E0=E; U0=v; t=thickness; Gam=r
P=global load
GK=Global stiffness matrix
NR=Number of row in global matrix for which the displacement is known
NL=Number of loads need to input
NDK=Number of Displacement Known
NLR=number of row in global for a load

Assignments:

Write a triangular finite element code independently.

Requirement:

(1) Students should not fully or partially copy the program from somebody else.

(2) Students should make the mesh automatically, not by hands. Chapters 13 and 14 introduce how to design the mesh automatically.

(3) Students should present the calculation results by contour plots.

(4) Students should make presentation to show how the finite element code was designed.

Chapter 7 Area coordinates and more node element

More node element refers to the element nodes more than three, such as six-node triangle element and four-node rectangle element, which will be studied. Before learning six-node triangle element, we first learn area coordinates which will be used to express the shape function of six-node triangle elements.

7.1 Area coordinates

Area coordinates definition: for a point P inside the triangular element as shown in Fig. 7—1, the area coordinates of this point are L_i, L_j and L_k, where

$$L_i = \frac{A_i}{A}, \quad L_j = \frac{A_j}{A}, \quad L_k = \frac{A_k}{A} \tag{7—1}$$

where A is the area of a triangular element, A_i, A_j and A_k are the areas of triangle P_{jk}, P_{ki} and P_{ij}, as showing in Fig. 7—1.

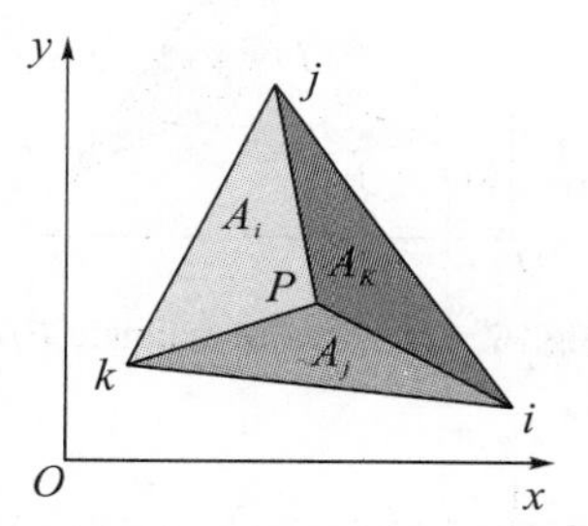

Fig. 7—1 Area coordinates

7.1.1 The relationship of area coordinates versus shape functions

Because $A_i + A_j + A_k = A$, we have

$$L_i + L_j + L_k = \frac{A_i}{A} + \frac{A_j}{A} + \frac{A_k}{A} = \frac{A}{A} = 1 \tag{7—2}$$

From Eq. (7—1) and Fig. 7—1, we can find as P coincides with a node, the corresponding area coordinate is 1, and the rest coordinates are 0, i. e.

$$\begin{aligned}&\text{At node } i: \quad L_i = 1,\ L_j = L_k = 0\\&\text{At node } j: \quad L_j = 1,\ L_i = L_k = 0\\&\text{At node } k: \quad L_k = 1,\ L_j = L_i = 0\end{aligned} \tag{7-3}$$

In Chapter 1, we have learnt that the shape functions have the similar properties, i. e. for a triangular element, we can have

$$N_i + N_j + N_k = 1$$

where $N_i = \frac{1}{2A}(a_i + b_i x + c_i y)$, and

$$\begin{aligned}&\text{At node } i: N_i = 1, N_j = N_k = 0\\&\text{At node } j: N_i = 0, N_j = 1, N_k = 0\\&\text{At node } k: N_i = 0, N_j = 0, N_k = 1\end{aligned}$$

Comparing the properties of the area coordinates versus those of the shape functions, one may guess that the area coordinates are just the shape functions, i. e. $L_i = N_i, L_j = N_j, L_k = N_k$.

Next, we will try to find the relation between area coordinates and the shape functions. Let's recall the calculation method of triangle area presented in Chapter 1.

$$D = \begin{vmatrix} 1 & x_i & y_i \\ 1 & x_j & y_j \\ 1 & x_k & y_k \end{vmatrix} = y_k(x_j - x_i) + y_j(x_i - x_k) + y_i(x_k - x_j) = 2A$$

[shown in Eq. (1−4)]

According to Eq. (1−4) and Fig. 7−2, the triangle Pjk area A_i can be obtained by

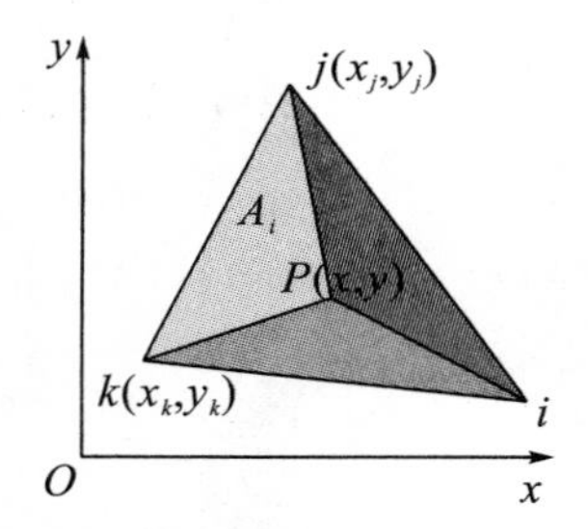

Fig. 7−2 Area of triangle *Pjk*

$$A_i = \frac{1}{2}\begin{vmatrix} 1 & x & y \\ 1 & x_j & y_j \\ 1 & x_k & y_k \end{vmatrix} = \frac{1}{2}[y_k(x_j - x) + y_j(x - x_k) + y(x_k - x_j)]$$

$$= \frac{1}{2}[x_j y_k - x_k y_j + (y_j - y_k)x + (x_k - x_j)y] \tag{7-4}$$

Because $a_i = x_j y_k - x_k y_j, b_i = y_j - y_k, c_i = -x_j + x_k$, we have

$$A_i = \frac{1}{2}(a_i + b_i x + c_i y) \tag{7-5}$$

According to Eq. (7−1), we have

$$L_i = \frac{A_i}{A} = \frac{1}{2A}(a_i + b_i x + c_i y) \tag{7-6}$$

Therefore, $L_i = N_i$, and similarly, we have

$$\begin{cases} L_i = \frac{1}{2A}(a_i + b_i x + c_i y) = N_i \\ L_j = \frac{1}{2A}(a_j + b_j x + c_j y) = N_j \\ L_k = \frac{1}{2A}(a_k + b_k x + c_k y) = N_k \end{cases} \tag{7-7}$$

7.1.2 The relationship of area coordinates versus Cartesian coordinates

Multiplying the area coordinates by x_i, x_j, x_k, respectively, we have

$$\begin{cases} x_i L_i = \frac{1}{2A}(a_i + b_i x + c_i y)x_i \\ x_j L_j = \frac{1}{2A}(a_j + b_j x + c_j y)x_j \\ x_k L_k = \frac{1}{2A}(a_k + b_k x + c_k y)x_k \end{cases} \tag{7-8}$$

One can prove that the sum of $x_i L_i$, $x_j L_j$ and $x_k L_k$ equals x, i. e.

$$x = x_i L_i + x_j L_j + x_k L_k \tag{7-9}$$

Similarly for the coordinate y, and finally we have

$$\begin{cases} x = x_i L_i + x_j L_j + x_k L_k \\ y = y_i L_i + y_j L_j + y_k L_k \end{cases} \tag{7-10}$$

Eq. (7—10) can be written in another form as

$$\begin{cases} x = N_i x_i + N_j x_j + N_k x_k \\ y = N_i y_i + N_j y_j + N_k y_k \end{cases} \tag{7-11}$$

Eq. (7—11) indicates that for three node triangle element, the coordinates x and y of a point in the triangle can be expressed by the sum of the shape function times the three node coordinates. In Chapter 1, we have learnt that the displacements have the similar properties, i. e.

$$\begin{cases} u = N_i u_i + N_j u_j + N_k u_k \\ v = N_i v_i + N_j v_j + N_k v_k \end{cases} \tag{7-12}$$

7.2 Selection method of general displacement function

Displacement function should be selected according to the Pascal (the French mathematician Blaise Pascal) triangle principle as shown in Fig. 7—3:

(1) The shape functions about the coordinates x and y must be symmetric, which means the coordinates x and y must have the exactly same forms in the shape functions,

such as the shape functions for triangle element lin Eq. (1—1), i. e. $\begin{cases} u = \alpha_1 + \alpha_2 x + \alpha_3 y \\ v = \alpha_4 + \alpha_5 x + \alpha_6 y \end{cases}$.

$$
\begin{array}{ccccccccc}
 & & & x & & y & & & \\
 & & x^2 & & xy & & y^2 & & \\
 & x^3 & & x^2y & & xy^2 & & y^3 & \\
x^4 & & x^3y & & x^2y^2 & & xy^3 & & y^4
\end{array}
$$

Fig. 7—3 Pascal triangle

(2) Shape functions must contain the low order terms, i. e. the shape function must contain constant and one order terms, and then if possible select the second order and third order terms. For example, the following two shape function are not allowed.

$u(x,y) = Ax^2 + Bx + Cy + D$ ⊗ the reason: it is not symmetric

$u(x,y) = Ax^2 + Bxy + Cy^2$ ⊗ the reason: it is no constant

The correct displacement function should be

$$u(x,y) = A + Bx + Cy + Dx^2 + Exy + Fy^2$$

7.3 Six-node triangular element

For the six-node triangle element shown in Fig. 7—4, there are six nodes and 12 freedoms (12 coefficients can be solved). Therefore, according to Pascal triangle principle, the displacement function can be selected as

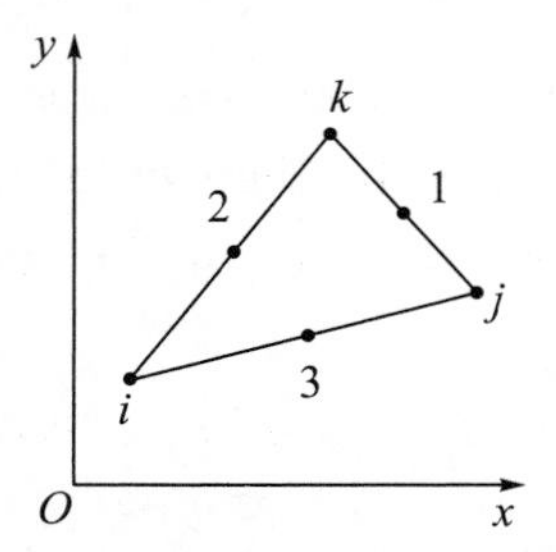

Fig. 7—4 Six-node triangle element

$$\begin{cases} u = \alpha_1 + \alpha_2 x + \alpha_3 y + \alpha_4 x^2 + \alpha_5 xy + \alpha_6 y^2 \\ v = \alpha_7 + \alpha_8 x + \alpha_9 y + \alpha_{10} x^2 + \alpha_{11} xy + \alpha_{12} y^2 \end{cases} \tag{7—13}$$

Similar as we did for the three node triangle element, the 12 coefficients can be obtained because there are 12 equations. After the 12 coefficients have been obtained, we substitute them into Eq. (7—13), and then the displacement functions can be written as

$$\begin{cases} u = N_i u_i + N_j u_j + N_k u_k + N_1 u_1 + N_2 u_2 + N_3 u_3 \\ v = N_i v_i + N_j v_j + N_k v_k + N_1 v_1 + N_2 v_2 + N_3 v_3 \end{cases} \tag{7—14}$$

The shape functions are related to the triangle area coordinates, and they can be written as

$$\begin{cases} N_i = L_i(2L_i - 1) \\ N_j = L_j(2L_j - 1) \\ N_k = L_k(2L_k - 1) \\ N_1 = 4L_jL_k \\ N_2 = 4L_iL_k \\ N_3 = 4L_iL_j \end{cases} \tag{7-15}$$

where $L_i = \dfrac{A_i}{A}$, $L_j = \dfrac{A_j}{A}$, $L_k = \dfrac{A_k}{A}$; and where A_i, A_j and A_k are shown in Fig. 7-5.

Eq. (7-15) can be simplified as

$$\begin{cases} N_i = L_i(2L_i - 1) & (i, j, k) \\ N_i = 4N_jN_k & (1, 2, 3) \end{cases} \tag{7-16}$$

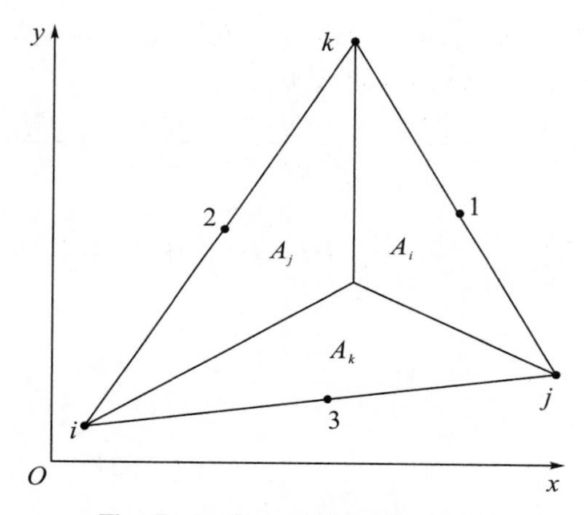

Fig. 7-5 Six-node triangle element

According to area coordinate definition, at nodes i, j and k the area coordinates (L_i, L_j, L_k) are i (1, 0, 0), j (0, 1, 0), k (0, 0, 1). At nodes 1, 2, 3 are (0, 0.5, 0.5), (0.5, 0, 0.5), (0.5, 0.5, 0). Therefore, at node i, $N_i = L_i(2L_i - 1) = 1$ due to $L_i = 1$; and at node 1, $N_1 = 4L_jL_k = 4 \times 0.5 \times 0.5 = 1$ due to $L_j = L_k = 0.5$, which means at each node, the corresponding shape function equals 1.

7.3.1 Strain matrix

Similar as three node triangle element, the strains can be calculated by the displacements as

$$\{\varepsilon\} = \begin{Bmatrix} \varepsilon_x \\ \varepsilon_y \\ \gamma_{xy} \end{Bmatrix} = \begin{Bmatrix} \dfrac{\partial u}{\partial x} \\ \dfrac{\partial v}{\partial y} \\ \dfrac{\partial u}{\partial y} + \dfrac{\partial v}{\partial x} \end{Bmatrix} = \begin{bmatrix} \dfrac{\partial}{\partial x} & 0 \\ 0 & \dfrac{\partial}{\partial y} \\ \dfrac{\partial}{\partial y} & \dfrac{\partial}{\partial x} \end{bmatrix} \begin{Bmatrix} u \\ v \end{Bmatrix} \tag{7-17}$$

Substituting the displacements in Eq. (7-14) into Eq. (7-17), we can have

$$\varepsilon_x = \frac{\partial u}{\partial x} = \frac{\partial N_i}{\partial x}u_i + \frac{\partial N_j}{\partial x}u_j + \frac{\partial N_k}{\partial x}u_k + \frac{\partial N_1}{\partial x}u_1 + \frac{\partial N_2}{\partial x}u_2 + \frac{\partial N_3}{\partial x}u_3 \tag{7-18}$$

where

$$\frac{\partial N_i}{\partial x} = \frac{\partial N_i}{\partial L_i}\frac{\partial L_i}{\partial x} = \frac{\partial [L_i(2L_i - 1)]}{\partial L_i}\frac{\partial \frac{1}{2A}(a_i + b_i x + c_i y)}{\partial x} = \frac{b_i(4L_i - 1)}{2A} \quad (i,j,k) \tag{7-19}$$

and

$$\frac{\partial N_1}{\partial x} = \frac{\partial N_1}{\partial L_j}\frac{\partial L_j}{\partial x} + \frac{\partial N_1}{\partial L_k}\frac{\partial L_k}{\partial x} = \frac{4(b_j L_k + b_k L_j)}{2A} \quad (1,2,3) \tag{7-20}$$

Substituting Eqs. (7—19) and (7—20) into Eq. (7—18), we have

$$\varepsilon_x = \frac{1}{2A}[b_i(4L_i - 1)u_i + b_j(4L_j - 1)u_j + b_k(4L_k - 1)u_k + 4(b_j L_k + b_k L_j)u_1 + 4(b_k L_i + b_i L_k)u_2 + 4(b_i L_j + b_j L_i)u_3] \tag{7-21}$$

Similarly, the strain ε_y and γ_{xy} can be obtained by

$$\varepsilon_y = \frac{\partial v}{\partial y} = \frac{\partial N_i}{\partial y}v_i + \frac{\partial N_j}{\partial y}v_j + \frac{\partial N_k}{\partial y}v_k + \frac{\partial N_1}{\partial y}v_1 + \frac{\partial N_2}{\partial y}v_2 + \frac{\partial N_3}{\partial y}v_3, \quad \gamma_{xy} = \frac{\partial u}{\partial y} + \frac{\partial v}{\partial x}$$

Finally

$$\{\varepsilon\} = [B]\{\delta\}^e = [[B_i][B_j][B_k][B_1][B_2][B_3]]\{\delta\}^e$$

and

$$[B] = \frac{1}{2A}\begin{bmatrix} b_i(4L_i-1) & 0 & b_j(4L_j-1) & 0 & b_k(4L_k-1) & 0 & 4(b_jL_k+b_kL_j) & 0 & 4(b_kL_i+b_iL_k) & 0 & 4(b_iL_j+b_jL_i) & 0 \\ 0 & c_i(4L_i-1) & 0 & c_j(4L_j-1) & 0 & c_k(4L_k-1) & 0 & 4(c_jL_k+c_kL_j) & 0 & 4(c_kL_i+c_iL_k) & 0 & 4(c_iL_j+c_jL_i) \\ c_i(4L_i-1) & b_i(4L_i-1) & c_j(4L_j-1) & b_j(4L_j-1) & c_k(4L_k-1) & b_k(4L_k-1) & 4(c_jL_k+c_kL_j) & 4(b_jL_k+b_kL_j) & 4(c_kL_i+c_iL_k) & 4(b_kL_i+b_iL_k) & 4(c_iL_j+c_jL_i) & 4(b_iL_j+b_jL_i) \end{bmatrix} \tag{7-22}$$

where

$$[B_i] = \frac{1}{2A}\begin{bmatrix} b_i(4L_i - 1) & 0 \\ 0 & c_i(4L_i - 1) \\ c_i(4L_i - 1) & b_i(4L_i - 1) \end{bmatrix} \quad (i,j,k)$$

$$[B_1] = \frac{1}{2A}\begin{bmatrix} 4(b_j L_k + b_k L_j) & 0 \\ 0 & 4(c_j L_k + c_k L_j) \\ 4(c_j L_k + c_k L_j) & 4(b_j L_k + b_k L_j) \end{bmatrix} \quad (1,2,3)$$

7.3.2 Stress matrix

Similar as three node triangle element, the stress can be written as

$$\{\sigma\} = [D]\{\varepsilon\} = [D][B]\{\delta\}^e = [S]\{\delta\}^e \tag{7-23}$$

Substituting $[B]$ in Eq. (7—22) and $[D]$ in Eq. (1—16) into Eq. (7—23), one can have

$$\{\sigma\} = [S]\{\delta\}^e = [[S_i][S_j][S_k][S_1][S_2][S_3]]\{\delta\}^e \tag{7-24}$$

where

$$[S_i] = \frac{Et(4L_i - 1)}{4(1-\nu^2)A}\begin{bmatrix} 2b_i & 2\nu c_i \\ 2\nu b_i & 2c_i \\ (1-\nu)c_i & (1-\nu)b_i \end{bmatrix} \quad (i,j,k)$$

$$[S_1]=\frac{Et}{4(1-\nu^2)A}\begin{bmatrix} 8(b_jL_k+b_kL_j) & 8\nu(c_jL_k+c_kL_j) \\ 8\nu(b_jL_k+b_kL_j) & 8(c_jL_k+c_kL_j) \\ 4(1-\nu)(c_jL_k+c_kL_j) & 4(1-\nu)(b_jL_k+b_kL_j) \end{bmatrix} \quad (1,2,3)$$

It can be seen that the matrix $[S]$ is the function of area coordinates, which is linearly related to the x, y coordinates. Therefore, the stresses of a six-node element are not constants any more, and they are linearly related to x and y coordinates.

7.3.3 Element stiffness matrix

The element stiffness matrix can be written as

$$[K]^e=\iint_A [B]^T[D][B]t\mathrm{d}x\mathrm{d}y=\iint_A [B]^T[S]t\mathrm{d}x\mathrm{d}y \quad (7-25)$$

where

$$[D]=\frac{E}{1-\nu^2}\begin{bmatrix} 1 & \nu & 0 \\ \nu & 1 & 0 \\ 0 & 0 & \frac{1-\nu}{2} \end{bmatrix} \quad \text{[shown in Eq. (1-16)]}$$

Substituting $[B]$ in Eq. (7-22) and $[D]$ in Eq. (1-16) into the above equation, we can find that the integrand contains the coordinates L_i, L_j and L_k, and it is not a constant as that for three node triangle element. The element stiffness can be expressed as

$$[K]^e=\begin{bmatrix} K_{ii} & K_{ij} & K_{ik} & K_{i1} & K_{i2} & K_{i3} \\ K_{ji} & K_{jj} & K_{jk} & K_{j1} & K_{j2} & K_{j3} \\ K_{ki} & K_{kj} & K_{kk} & K_{k1} & K_{k2} & K_{k3} \\ K_{1i} & K_{1j} & K_{1k} & K_{11} & K_{12} & K_{13} \\ K_{2i} & K_{2j} & K_{2k} & K_{21} & K_{22} & K_{23} \\ K_{3i} & K_{3j} & K_{3k} & K_{31} & K_{32} & K_{33} \end{bmatrix} \quad (7-26)$$

By using the following formula

$$\iint L_i^\alpha L_j^\beta L_k^\gamma \mathrm{d}x\mathrm{d}y=\frac{\alpha!\beta!\gamma!}{(\alpha+\beta+\gamma+2)!}2A \quad (7-27)$$

We can have

$$[K]^e=\frac{Et}{24(1-\nu^2)A}\begin{bmatrix} A_i & G_{ij} & G_{ik} & 0 & -4G_{ik} & -4G_{ij} \\ G_{ij} & A_j & G_{jk} & -4G_{jk} & 0 & -4G_{ji} \\ G_{ki} & G_{kj} & A_k & -4G_{kj} & -4G_{ki} & 0 \\ 0 & -4G_{kj} & -4G_{jk} & B_i & D_{ij} & D_{ik} \\ -4G_{ki} & 0 & -4G_{ik} & D_{ji} & B_j & D_{jk} \\ -4G_{ji} & -4G_{ij} & 0 & D_{ki} & D_{kj} & B_k \end{bmatrix} \quad (7-28)$$

where

$$[A_i]=\begin{bmatrix} 6b_i^2+3(1-\nu)c_i^2 & 3(1+\nu)b_ic_i \\ 3(1+\nu)b_ic_i & 6c_i^2+3(1-\nu)b_i^2 \end{bmatrix} \quad (i,j,k)$$

$$[B_i]=\begin{bmatrix}16(b_i^2-b_jb_k)+8(1-\nu)(c_i^2-c_jc_k) & 4(1+\nu)(b_ic_i+b_jc_j+b_kc_k)\\ 4(1+\nu)(b_ic_i+b_jc_j+b_kc_k) & 16(c_i^2-c_jc_k)+8(1-\nu)(b_i^2-b_jb_k)\end{bmatrix}$$

$$(i,j,k)$$

$$[G_{rs}]=\begin{bmatrix}-2b_rb_s-(1-\nu)c_rc_s & -2\nu b_rc_s-(1-\nu)c_rb_s\\ -\nu c_rb_s-(1-\nu)b_rc_s & -2c_rc_s-(1-\nu)b_rb_s\end{bmatrix}\quad (i,j,k)$$

$$[D_{rs}]=\begin{bmatrix}16b_rb_s+8(1-\nu)c_rc_s & 4(1+\nu)(c_rb_s+b_rc_s)\\ 4(1+\nu)(c_rb_s+b_rc_s) & 16c_rc_s+8(1-\nu)b_rb_s\end{bmatrix}\quad (i,j,k)$$

$$\begin{cases}b_i=y_j-y_k\\ c_i=-x_j+x_k\end{cases}\quad (i,j,k)$$

7.3.4 Global stiffness equation

Similar as for three node triangle element, the global stiffness matrix consists of element stiffness matrices. When we obtain an element stiffness matrix, we should install the element stiffness matrix into the global matrix according to the real node number of the element. For example, for the two six-node triangle elements shown in Fig. 7—6, the global equation can be written as

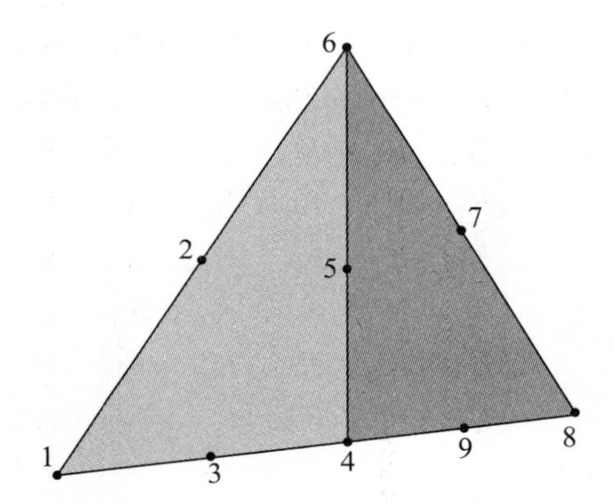

Fig. 7—6 Two six-node triangle element

$$\begin{bmatrix}K_{11} & K_{12} & K_{13} & K_{14} & K_{15} & K_{16} & & & \\ K_{21} & K_{22} & K_{23} & K_{24} & K_{25} & K_{26} & & & \\ K_{31} & K_{32} & K_{33} & K_{34} & K_{35} & K_{36} & & & \\ K_{41} & K_{42} & K_{43} & K_{44} & K_{45} & K_{46} & K_{47} & K_{48} & K_{49}\\ K_{51} & K_{52} & K_{53} & K_{54} & K_{55} & K_{56} & K_{57} & K_{58} & K_{59}\\ K_{61} & K_{62} & K_{63} & K_{64} & K_{65} & K_{66} & K_{67} & K_{68} & K_{69}\\ & & & K_{74} & K_{75} & K_{76} & K_{77} & K_{78} & K_{79}\\ & & & K_{84} & K_{85} & K_{86} & K_{87} & K_{88} & K_{89}\\ & & & K_{94} & K_{95} & K_{96} & K_{97} & K_{98} & K_{99}\end{bmatrix}\begin{Bmatrix}\delta_1\\ \delta_2\\ \delta_3\\ \delta_4\\ \delta_5\\ \delta_6\\ \delta_7\\ \delta_8\\ \delta_9\end{Bmatrix}=\begin{Bmatrix}F_1\\ F_2\\ F_3\\ F_4\\ F_5\\ F_6\\ F_7\\ F_8\\ F_9\end{Bmatrix}\qquad (7-29)$$

7.4 Four-node rectangle element

For the rectangle element shown in Fig. 7－7, there are 4 nodes and 8 freedoms (8 coefficients can be solved). Therefore, according to Pascal triangle principle, the displacement function can be selected as

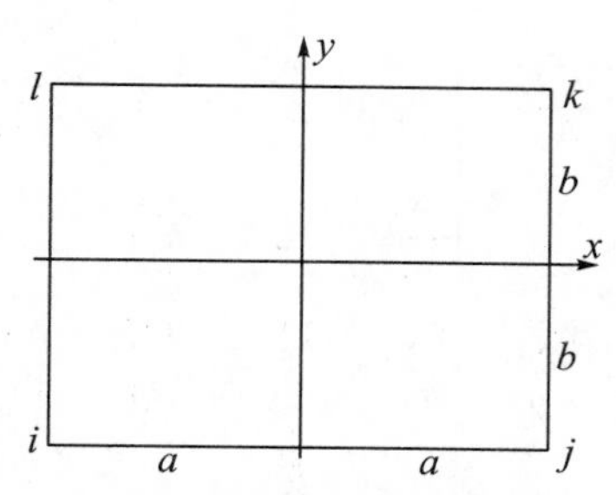

Fig. 7—7 Rectangle element

$$\begin{cases} u = \alpha_1 + \alpha_2 x + \alpha_3 y + \alpha_4 xy \\ v = \alpha_5 + \alpha_6 x + \alpha_7 y + \alpha_8 xy \end{cases} \tag{7-30}$$

Similar to the three-node triangle element, the 8 coefficients can be obtained due to 8 equations. After the 8 coefficients have been obtained, substitute them into Eq. (7—30), and then the displacement functions can be obtained

$$\begin{cases} u = N_i u_i + N_j u_j + N_k u_k + N_l u_l \\ v = N_i v_i + N_j v_j + N_k v_k + N_l v_l \end{cases} \tag{7-31}$$

where

$$\begin{cases} N_i = \frac{1}{4}(1 - \frac{x}{a})(1 - \frac{y}{b}) \\ N_j = \frac{1}{4}(1 + \frac{x}{a})(1 - \frac{y}{b}) \\ N_k = \frac{1}{4}(1 + \frac{x}{a})(1 + \frac{y}{b}) \\ N_l = \frac{1}{4}(1 - \frac{x}{a})(1 + \frac{y}{b}) \end{cases} \tag{7-32}$$

It is easy to find that $N_i + N_j + N_k + N_l = 1$. At node i, $N_i = 1$, $N_j = N_k = N_l = 0$, and similar for those at node j, k and l.

7.4.1 Strain matrix

According to relationship between strains and displacements, i.e. Eq. (7—17)

$$\{\varepsilon\}=\begin{Bmatrix}\varepsilon_x\\ \varepsilon_y\\ \gamma_{xy}\end{Bmatrix}=\begin{Bmatrix}\dfrac{\partial u}{\partial x}\\ \dfrac{\partial v}{\partial y}\\ \dfrac{\partial u}{\partial y}+\dfrac{\partial v}{\partial x}\end{Bmatrix}=\begin{bmatrix}\dfrac{\partial}{\partial x} & 0\\ 0 & \dfrac{\partial}{\partial y}\\ \dfrac{\partial}{\partial y} & \dfrac{\partial}{\partial x}\end{bmatrix}\begin{Bmatrix}u\\ v\end{Bmatrix}$$

[shown in Eq. (7−17)]

Substituting Eq. (7−31) into Eq. (7−17), we have

$$\{\varepsilon\}=\begin{Bmatrix}\varepsilon_x\\ \varepsilon_y\\ \gamma_{xy}\end{Bmatrix}=\begin{bmatrix}\dfrac{\partial u}{\partial x}\\ \dfrac{\partial v}{\partial y}\\ \dfrac{\partial u}{\partial y}+\dfrac{\partial v}{\partial x}\end{bmatrix}=\left[\begin{array}{cc:cc:cc:cc}\dfrac{\partial N_i}{\partial x} & 0 & \dfrac{\partial N_j}{\partial x} & 0 & \dfrac{\partial N_k}{\partial x} & 0 & \dfrac{\partial N_l}{\partial x} & 0\\ 0 & \dfrac{\partial N_i}{\partial y} & 0 & \dfrac{\partial N_j}{\partial y} & 0 & \dfrac{\partial N_k}{\partial y} & 0 & \dfrac{\partial N_l}{\partial y}\\ \dfrac{\partial N_i}{\partial y} & \dfrac{\partial N_i}{\partial x} & \dfrac{\partial N_j}{\partial y} & \dfrac{\partial N_j}{\partial x} & \dfrac{\partial N_k}{\partial y} & \dfrac{\partial N_k}{\partial x} & \dfrac{\partial N_l}{\partial y} & \dfrac{\partial N_l}{\partial x}\end{array}\right]\begin{Bmatrix}u_i\\ v_i\\ u_j\\ v_j\\ u_k\\ v_k\\ u_l\\ v_l\end{Bmatrix} \tag{7−33}$$

Eq. (7−33) can be simplied as

$$\{\varepsilon\}=[B]\{\delta\}^e=[B_i\ B_j\ B_k\ B_l]\{\delta\}^e \tag{7−34}$$

where

$$[B]=\begin{bmatrix}\dfrac{\partial N_i}{\partial x} & 0 & \dfrac{\partial N_j}{\partial x} & 0 & \dfrac{\partial N_k}{\partial x} & 0 & \dfrac{\partial N_l}{\partial x} & 0\\ 0 & \dfrac{\partial N_i}{\partial y} & 0 & \dfrac{\partial N_j}{\partial y} & 0 & \dfrac{\partial N_k}{\partial y} & 0 & \dfrac{\partial N_l}{\partial y}\\ \dfrac{\partial N_i}{\partial y} & \dfrac{\partial N_i}{\partial x} & \dfrac{\partial N_j}{\partial y} & \dfrac{\partial N_j}{\partial x} & \dfrac{\partial N_k}{\partial y} & \dfrac{\partial N_k}{\partial x} & \dfrac{\partial N_l}{\partial y} & \dfrac{\partial N_l}{\partial x}\end{bmatrix}$$

Substituting the shape functions in Eq. (7−32) into the $[B]$ matrix, we have

$$[B]=\frac{1}{4}\left[\begin{array}{cc:cc:cc:cc}-\frac{1}{a}(1-\frac{y}{b}) & 0 & \frac{1}{a}(1-\frac{y}{b}) & 0 & \frac{1}{a}(1+\frac{y}{b}) & 0 & -\frac{1}{a}(1+\frac{y}{b}) & 0\\ 0 & -\frac{1}{b}(1-\frac{x}{a}) & 0 & -\frac{1}{b}(1+\frac{x}{a}) & 0 & \frac{1}{b}(1+\frac{x}{a}) & 0 & \frac{1}{b}(1-\frac{x}{a})\\ -\frac{1}{b}(1-\frac{x}{a}) & -\frac{1}{a}(1-\frac{y}{b}) & -\frac{1}{b}(1+\frac{x}{a}) & \frac{1}{a}(1-\frac{y}{b}) & \frac{1}{b}(1+\frac{x}{a}) & \frac{1}{a}(1+\frac{y}{b}) & \frac{1}{b}(1-\frac{x}{a}) & -\frac{1}{a}(1+\frac{y}{b})\end{array}\right] \tag{7−35}$$

7.4.2 Stress matrix

According to the relationship between stresses and strains, i. e.

$$\{\sigma\}=[D]\{\varepsilon\}=[D][B]\{\delta\}^e=[S]\{\delta\}^e \tag{7−36}$$

$$[D]=\frac{E}{1-\nu^2}\begin{bmatrix}1 & \nu & 0\\ \nu & 1 & 0\\ 0 & 0 & \dfrac{1-\nu}{2}\end{bmatrix}$$

[shown in Eq. (1−16)]

Substituting $[B]$ in Eq. (7−35) and $[D]$ in Eq. (1−16) into Eq. (7−36), we can get the $[S]$ matrix.

$$[S]=\frac{E}{4(1-\nu^2)}\begin{bmatrix} -\frac{1}{a}(1-\frac{y}{b}) & -\frac{\nu}{b}(1-\frac{x}{a}) & \frac{1}{a}(1-\frac{y}{b}) & -\frac{\nu}{b}(1+\frac{x}{a}) & \frac{1}{a}(1+\frac{y}{b}) & \frac{\nu}{b}(1+\frac{x}{a}) & -\frac{1}{a}(1+\frac{y}{b}) & \frac{\nu}{b}(1-\frac{x}{a}) \\ -\frac{\nu}{a}(1-\frac{y}{b}) & -\frac{1}{b}(1-\frac{x}{a}) & \frac{\nu}{a}(1-\frac{y}{b}) & -\frac{1}{b}(1+\frac{x}{a}) & \frac{\nu}{a}(1+\frac{y}{b}) & \frac{1}{b}(1+\frac{x}{a}) & -\frac{\nu}{a}(1+\frac{y}{b}) & \frac{1}{b}(1-\frac{x}{a}) \\ -\frac{1-\nu}{2b}(1-\frac{x}{a}) & -\frac{1-\nu}{2a}(1-\frac{y}{b}) & -\frac{1-\nu}{2b}(1+\frac{x}{a}) & \frac{1-\nu}{2a}(1-\frac{y}{b}) & \frac{1-\nu}{2b}(1+\frac{x}{a}) & \frac{1-\nu}{2a}(1+\frac{y}{b}) & \frac{1-\nu}{2b}(1-\frac{x}{a}) & -\frac{1-\nu}{2a}(1+\frac{y}{b}) \end{bmatrix} \tag{7-37}$$

7.4.3 Element stiffness matrix

Similar as three node triangle element, by using Principle of virtual displacement, we can get the element stiffness as

$$[K]^e=\iint_A [B]^{\mathrm{T}}[D][B]t\,\mathrm{d}x\mathrm{d}y=\iint_A [B]^{\mathrm{T}}[S]t\,\mathrm{d}x\mathrm{d}y \tag{7-38}$$

Both the $[B]$ matrix in Eq. (7−35) and $[S]$ matrix in Eq. (7−37) contain the coordinates x and y, thus the integrand of Eq. (7−38) cannot be moved out as that for three-node triangle element. Under such scenario, we have to use Gauss integration method to solve it, which we will learn later on, Because $[B]^{\mathrm{T}}$ is a 8×3 matrix and $[S]$ is a 3×8 matrix, the element stiffness must be a 8×8 matrix, and it can be written as

$$[K]^e=\begin{bmatrix} K_{ii} & K_{ij} & K_{ik} & K_{il} \\ K_{ji} & K_{jj} & K_{jk} & K_{jl} \\ K_{ki} & K_{kj} & K_{kk} & K_{kl} \\ K_{li} & K_{lj} & K_{lk} & K_{ll} \end{bmatrix} \tag{7-39}$$

7.4.4 Global stiffness equation for rectangle element

Similar as for three node triangle element, the global stiffness matrix consists of the element stiffness matrices. For each element, the stiffness matrix should be installed into the global matrix according to the real node number of the element. For example, for the three rectangle elements shown in Fig. 7−8, the global equation can be written as

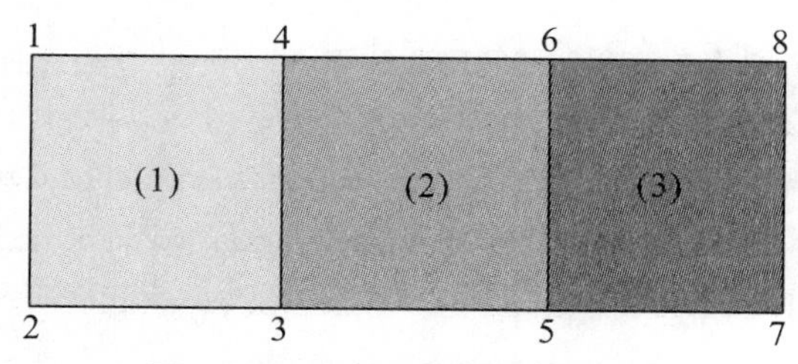

Fig. 7−8 Three rectangle elements

$$\begin{bmatrix} K_{11} & K_{12} & K_{13} & K_{14} & & & & \\ K_{21} & K_{22} & K_{23} & K_{24} & & & & \\ K_{31} & K_{32} & K_{33} & K_{34} & K_{35} & K_{36} & & \\ K_{41} & K_{42} & K_{43} & K_{44} & K_{45} & K_{46} & & \\ & & K_{53} & K_{54} & K_{55} & K_{56} & K_{57} & K_{58} \\ & & K_{63} & K_{64} & K_{65} & K_{66} & K_{67} & K_{68} \\ & & & & K_{75} & K_{76} & K_{77} & K_{78} \\ & & & & K_{85} & K_{86} & K_{87} & K_{88} \\ & & & & K_{95} & K_{96} & K_{97} & K_{98} \end{bmatrix} \begin{Bmatrix} \delta_1 \\ \delta_2 \\ \delta_3 \\ \delta_4 \\ \delta_5 \\ \delta_6 \\ \delta_7 \\ \delta_8 \\ \delta_9 \end{Bmatrix} = \begin{Bmatrix} F_1 \\ F_2 \\ F_3 \\ F_4 \\ F_5 \\ F_6 \\ F_7 \\ F_8 \\ F_9 \end{Bmatrix} \qquad (7-40)$$

7.4.5 Discussion about the rectangle element

For the rectangle element, the displacement functions are

$$\begin{cases} u = \alpha_1 + \alpha_2 x + \alpha_3 y + \alpha_4 xy \\ v = \alpha_5 + \alpha_6 x + \alpha_7 y + \alpha_8 xy \end{cases} \qquad \text{[shown in Eq. (7-30)]}$$

One can find that the displacements are not linearly related to the x and y coordinates. At the mutual boundary, as shown in Fig. 7-9, the deformation compatibility may not be satisfied.

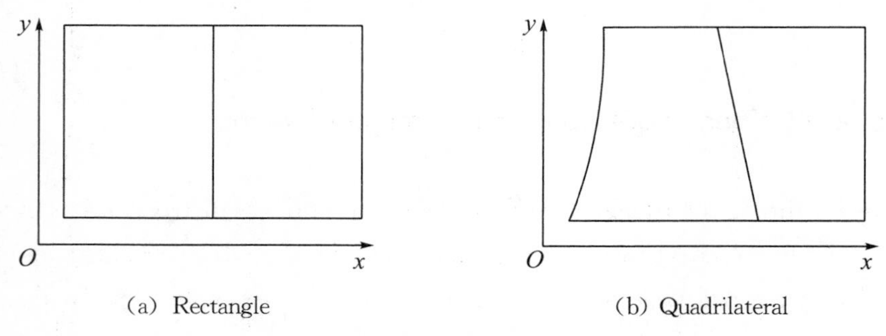

(a) Rectangle (b) Quadrilateral

Fig. 7-9 Two rectangle elements

As is well known, deformation compatibility is the key issue for studying material mechanics behavior (stress and strain), and we must obey this rule. If the displacement function cannot satisfy the deformation compatibility, the displacements at the mutual boundary are not continuous, and the dislocation of gap or overlap may occur, which means the rectangle element method is not effective. However, because each edge of rectangle elements in Fig. 7-9 (a) parallel to either the x-axis or y-axis, which makes either x or y in Eq. (7-30) be a constant for each edge of rectangle elements. Therefore, the displacement functions in Eq. (7-30) actually are linearly related to the x or y coordinate, and they can satisfy the deformation compatibility.

For non-rectangle four-node element or quadrilateral element shown in Fig. 7-9 (b), because the displacement at the mutual boundary is not linear, the deformation compatibility cannot be satisfied. Therefore, non-rectangle four-node elements generally cannot be directly applied in the finite element method because they may have inclined edges or curved edges. For solving such problem, we will learn isoparametric element method later on, in which the element could be any shaped.

Assignments:

1. A six-node triangular element is shown in the following figure. The edge 142 is subjected to a uniformly distributed pressure q, and the thickness of the plate is t. Calculate the equivalent nodal forces.

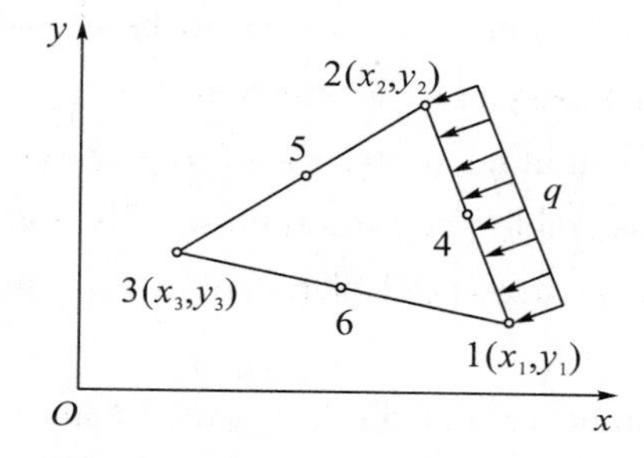

2. When using four-node elements, the element must be rectangle or square. Explain the reason.

3. Show the relationship between the area coordinates and the shape functions for a three-node triangle element.

Chapter 8　Axisymmetric stress analysis

In the axisymmetrical situation, any radial displacement automatically induces a strain in the circumferential direction, and as the stresses in this direction are certainly non-zero, this fourth component of strain and of the associated stress has to be considered. Here lies the essential difference in the treatment of the axisymmetric situation.

The problem of stress distribution in bodies of revolution (axisymmetric solids) under axisymmetric loading is of considerable practical interest. The mathematical problems presented are very similar to those of plane stress and plane strain as, once again, the situation is two dimensional.

By symmetry, the two components of displacements in any plane section of the body along its axis of symmetry define completely the state of strain and, therefore, the state of stress. Such a cross-section is shown in this Fig. 8—1.

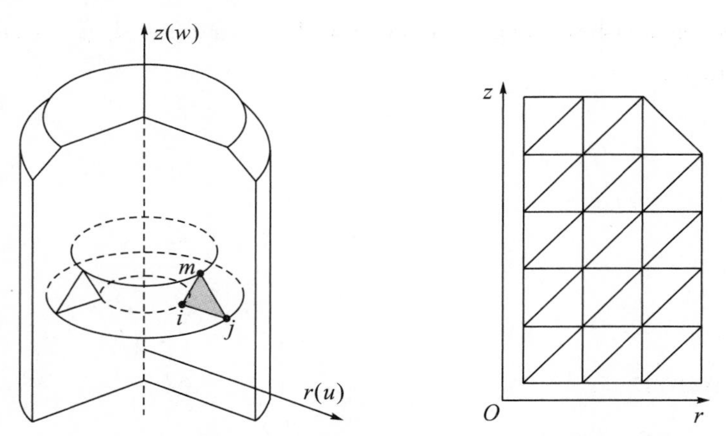

Fig. 8—1　An axisymmetrical structure and its cross-section plane

The volume of material associated with an 'element' is now that of a body of revolution. The triangular element is again used mainly for illustrative purpose, the principal developed being completely general. The algebra here may be somehow more tedious than that in the previous chapter, but essentially, identical operations will be involved. We arbitrarily select one triangle element, as shown in Fig. 8—2.

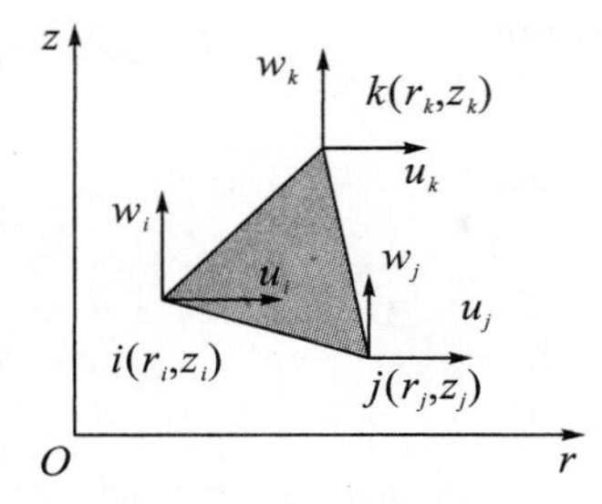

Fig. 8−2 Three node element

Similar to the three-node element, the displacement functions can be written as

$$\begin{cases} u = \alpha_1 + \alpha_2 r + \alpha_3 z \\ w = \alpha_4 + \alpha_5 r + \alpha_6 z \end{cases} \tag{8-1}$$

And the relationship of displacements of a point versus nodal displacements can be expressed as

$$\begin{Bmatrix} u \\ w \end{Bmatrix} = \begin{Bmatrix} N_i u_i + N_j u_j + N_k u_k \\ N_i w_i + N_j w_j + N_k w_k \end{Bmatrix} = \begin{bmatrix} N_i & 0 & N_j & 0 & N_k & 0 \\ 0 & N_i & 0 & N_j & 0 & N_k \end{bmatrix} \begin{Bmatrix} u_i \\ w_i \\ u_j \\ w_j \\ u_k \\ w_k \end{Bmatrix} = [N]\{\delta\}^e \tag{8-2}$$

where

$$N_i = \frac{1}{2A}(a_i + b_i r + c_i z) \quad (i,j,k) \tag{8-3}$$

and

$$\begin{cases} a_i = r_j z_k - r_k z_j \\ b_i = z_j - z_k \\ c_i = r_k - r_j \end{cases} \quad (i,j,k) \tag{8-4}$$

8.1 Strain matrix

For axisymmetrical problem, any radial displacement automatically induces a strain and stress in the circumferential direction, thus comparing two-dimensional issue, one extra component of strain and of the associated stress exist. Usually the fourth component of stress plays a key role in material fracturing behavior, such as water pipe failure in winter in north area. Therefore, the fourth component of stress and strain has to be considered. The relationship between strains and displacement can be expressed as

$$\{\varepsilon\}=\begin{Bmatrix}\varepsilon_r\\ \varepsilon_z\\ \gamma_{rz}\\ \varepsilon_\theta\end{Bmatrix}=\begin{Bmatrix}\dfrac{\partial u}{\partial r}\\ \dfrac{\partial w}{\partial z}\\ \dfrac{\partial u}{\partial z}+\dfrac{\partial w}{\partial r}\\ \dfrac{u}{r}\end{Bmatrix}=\begin{bmatrix}\dfrac{\partial}{\partial r} & 0\\ 0 & \dfrac{\partial}{\partial z}\\ \dfrac{\partial}{\partial z} & \dfrac{\partial}{\partial r}\\ \dfrac{1}{r} & 0\end{bmatrix}\begin{Bmatrix}u\\ w\end{Bmatrix} \tag{8-5}$$

Substituting the displacement from Eq. (8−2) into Eq. (8−5), one can have

$$\{\varepsilon\}=\begin{Bmatrix}\varepsilon_r\\ \varepsilon_z\\ \gamma_{rz}\\ \varepsilon_\theta\end{Bmatrix}=[B]\{\delta\}^e=\frac{1}{2A}\begin{bmatrix}b_i & 0 & b_j & 0 & b_k & 0\\ 0 & c_i & 0 & c_j & 0 & c_k\\ c_i & b_i & c_j & b_j & c_k & b_k\\ f_i & 0 & f_j & 0 & f_k & 0\end{bmatrix}\begin{Bmatrix}u_i\\ w_i\\ u_j\\ w_j\\ u_k\\ w_k\end{Bmatrix} \tag{8-6}$$

where $f_i=\frac{a_i}{r}+b_i+\frac{c_i z}{r}$, $a_i=r_j z_k-r_k z_j$, $b_i=z_j-z_k$, $c_i=r_k-r_j\,(i,j,k)$.

8.2 Stress matrix

For axial symmetric problem, the $[D]$ matrix is

$$[D]=\frac{E(1-\nu)}{(1+\nu)(1-2\nu)}\begin{bmatrix}1 & \dfrac{\nu}{1-\nu} & 0 & \dfrac{\nu}{1-\nu}\\ \dfrac{\nu}{1-\nu} & 1 & 0 & \dfrac{\nu}{1-\nu}\\ 0 & 0 & \dfrac{1-\nu}{2} & 0\\ \dfrac{\nu}{1-\nu} & \dfrac{\nu}{1-\nu} & 0 & 1\end{bmatrix} \tag{8-7}$$

The relationship between stresses and strain can be written as

$$\{\sigma\}=[D]\{\varepsilon\}=[D][B]\{\delta\}^e=[S]\{\delta\}^e \tag{8-8}$$

Substituting $[D]$ and $[B]$ matrix into Eq. (8−8), one can have

$$\begin{Bmatrix}\sigma_r\\ \sigma_z\\ \tau_{rz}\\ \sigma_\theta\end{Bmatrix}=\frac{E(1-\nu)}{2A(1+\nu)(1-2\nu)}\begin{bmatrix}b_i+A_1 f_i & A_1 c_i & b_j+A_1 f_j & A_1 c_j & b_k+A_1 f_k & A_1 c_k\\ A_1(b_i+f_i) & c_i & A_1(b_j+f_j) & c_j & A_1(b_k+f_k) & c_k\\ A_2 c_i & A_2 b_i & A_2 c_j & A_2 b_j & A_2 c_k & A_2 b_k\\ A_1 b_i+f_i & A_1 c_i & A_1 b_j+f_j & A_1 c_j & A_1 b_k+f_k & A_1 c_k\end{bmatrix}\begin{Bmatrix}u_i\\ w_i\\ u_j\\ w_j\\ u_k\\ w_k\end{Bmatrix} \tag{8-9}$$

where $A_1 = \frac{\nu}{1-\nu}$, $A_2 = \frac{1-2\nu}{2(1-\nu)}$, $f_i = \frac{a_i}{r} + b_i + \frac{c_i z}{r}$, $a_i = r_j z_k - r_k z_j$, $b_i = z_j - z_k$, $c_i = r_k - r_j (i, j, k)$.

It can be seen that except for the shear stress, all the normal stresses are not constants.

8.3 Elements stiffness matrix

For axisymmetrical problem, the element thickness t is different from those of plane problem, and it is the circle perimeter $2\pi r$. The whole element is a torus as shown in Fig. 8—3.

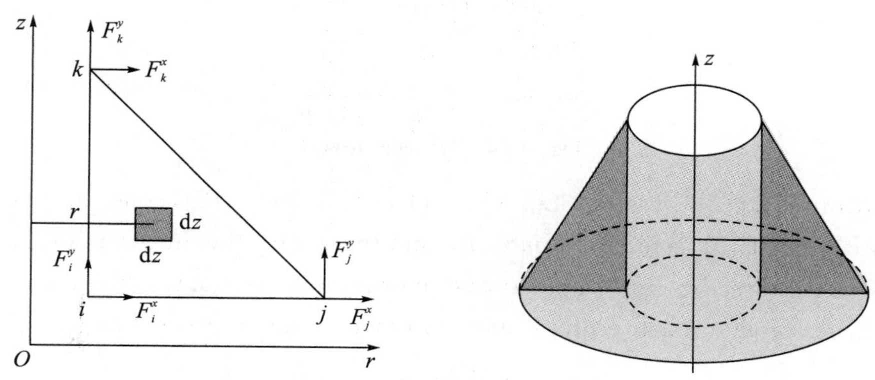

Fig. 8—3 Triangle element

Based on the principle of virtual displacement, we can have

$$(\{\delta^*\}^e)^T \{F\}^e = 2\pi \iint_A \{\varepsilon^*\}^T \{\sigma\} r \mathrm{d}r \mathrm{d}z = 2\pi \iint_A (\{\delta^*\}^e)^T [B]^T [D][B] \{\delta\}^e r \mathrm{d}r \mathrm{d}z \tag{8—10}$$

where $\{\varepsilon^*\} = [B]\{\delta^*\}^e$ and $\{\sigma\} = [D][B]\{\delta\}^e$.

The node displacements are constants, and they can be moved out from the integral. Eq. (8—10) can be rewritten as

$$\{F\}^e = 2\pi \iint_A [B]^T [D][B] r \mathrm{d}r \mathrm{d}z \cdot \{\delta\}^e = 2\pi \iint_A [B]^T [S] r \mathrm{d}r \mathrm{d}z \cdot \{\delta\}^e = [K]^e \{\delta\}^e \tag{8—11}$$

The element stiffness matrix can be written as

$$[K]^e = 2\pi \iint_A [B]^T [S] r \mathrm{d}r \mathrm{d}z \tag{8—12}$$

where

$$[B] = \frac{1}{2A} \begin{bmatrix} b_i & 0 & b_j & 0 & b_k & 0 \\ 0 & c_i & 0 & c_j & 0 & c_k \\ c_i & b_i & c_j & b_j & c_k & b_k \\ f_i & 0 & f_j & 0 & f_k & 0 \end{bmatrix} \tag{8—13}$$

$$[S] = \frac{E(1-\nu)}{2A(1+\nu)(1-2\nu)}\begin{bmatrix} b_i + A_1 f_i & A_1 c_i & b_j + A_1 f_j & A_1 c_j & b_k + A_1 f_k & A_1 c_k \\ A_1(b_i + f_i) & c_i & A_1(b_j + f_j) & c_j & A_1(b_k + f_k) & c_k \\ A_2 c_i & A_2 b_i & A_2 c_j & A_2 b_j & A_2 c_k & A_2 b_k \\ A_1 b_i + f_i & A_1 c_i & A_1 b_j + f_j & A_1 c_j & A_1 b_k + f_k & A_1 c_k \end{bmatrix} \tag{8-14}$$

where $A_1 = \frac{\nu}{1-\nu}$, $A_2 = \frac{1-2\nu}{2(1-\nu)}$, $f_i = \frac{a_i}{r} + b_i + \frac{c_i z}{r}$, $a_i = r_j z_k - r_k z_j$, $b_i = z_j - z_k$, $c_i = r_k - r_j (i, j, k)$.

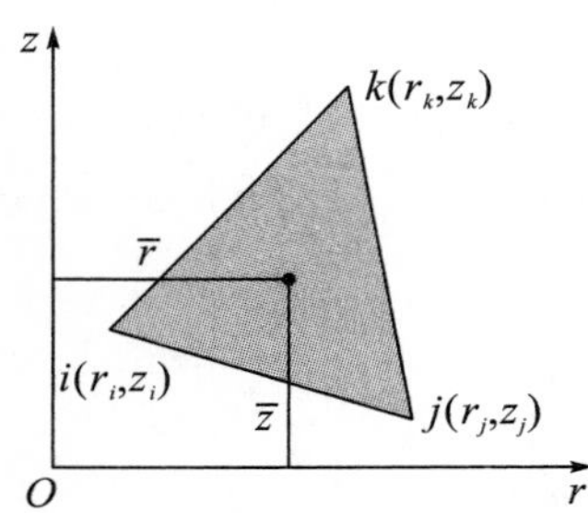

Fig. 8—4 Triangle element

Substituting Eq. (8—13) and Eq. (8—14) into Eq. (8—12), one can find that the element stiffness matrix contains variable r, and therefore, the integrand in Eq. (8—12) cannot move out directly. For simplicity and in order to avoid the trouble as $r=0$, we use the coordinates of an element center $\bar{r}$ and $\bar{z}$ to replace r and z, as shown in Fig. 8—4.

$$r \approx \bar{r} = \frac{1}{3}(r_i + r_j + r_k) \tag{8-15}$$

$$z \approx \bar{z} = \frac{1}{3}(z_i + z_j + z_k) \tag{8-16}$$

Accordingly, the parameter f_i can be expressed as

$$f_i \approx \bar{f}_i = \frac{a_i}{\bar{r}} + b_i + \frac{c_i \bar{z}}{\bar{r}} \quad (i, j, k) \tag{8-17}$$

After this operation, the coordinates of $\bar{r}$ and $\bar{z}$ are constant, and the strain matrix $[B]$ and stress matrix $[S]$ will change to constant matrices. The element stiffness matrix can be rewritten as

$$[K]^e = 2\pi \iint_A [B]^T[D][B] r\,dr\,dz = 2\pi\bar{r}\,[B]^T[D][B]A = \begin{bmatrix} K_{ii} & K_{ij} & K_{ik} \\ K_{ji} & K_{jj} & K_{jk} \\ K_{ki} & K_{kj} & K_{kk} \end{bmatrix} \tag{8-18}$$

where

$$[K_{rs}] = \frac{\pi E(1-\nu)\bar{r}}{2A(1+\nu)(1-2\nu)}\begin{bmatrix} b_r b_s + f_r f_s + A_1(b_r f_s + f_r b_s) + A_2 c_r c_s & A_1 c_s (b_r + f_r) + A_2 c_r b_s \\ A_1 c_r (b_s + f_s) + A_2 c_s b_r & c_r c_s + A_2 b_r b_s \end{bmatrix} \quad (r, s = i, j, k) \tag{8-19}$$

where $A_1 = \frac{\nu}{1-\nu}$, $A_2 = \frac{1-2\nu}{2(1-\nu)}$, $f_i \approx \bar{f}_i = \frac{a_i}{\bar{r}} + b_i + \frac{c_i \bar{z}}{\bar{r}}$, $a_i = r_j z_k - r_k z_j$, $b_i = z_j -$

$z_k, c_i = r_k - r_j (i, j, k)$.

The experience has shown that using the approximate method, which is easy, the results can satisfy the requirements of calculation precision as the mesh is condensed enough.

8.4 Equivalent nodal force

The forces for axisymmetrical problemmainly refer to body force and distributed force. No single concentrated force is allowed due to the symmetrical requirements (except for the structure, the load must be symmetrical either).

8.4.1 Body force

The body force is axisymmetric due to the symmetrical structure. For concentrated force in three node element, the equivalent node force can be expressed as

$$\{F\}^e = \begin{Bmatrix} F_i^x \\ F_i^y \\ F_j^x \\ F_j^y \\ F_k^x \\ F_k^y \end{Bmatrix} = \begin{Bmatrix} N_i & 0 \\ 0 & N_i \\ N_j & 0 \\ 0 & N_j \\ N_k & 0 \\ 0 & N_k \end{Bmatrix} \begin{Bmatrix} P_x \\ P_y \end{Bmatrix} \tag{8-20}$$

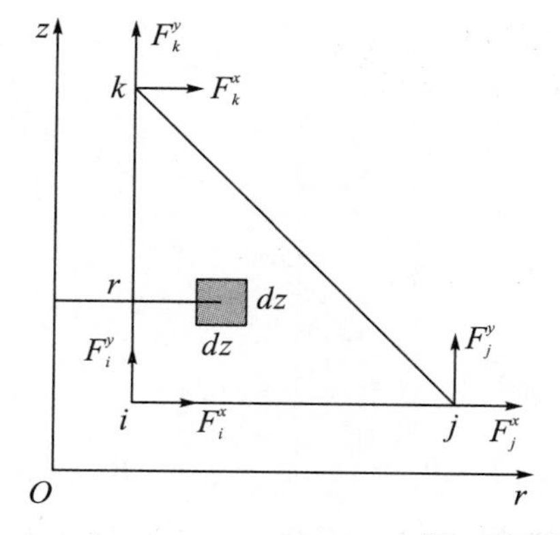

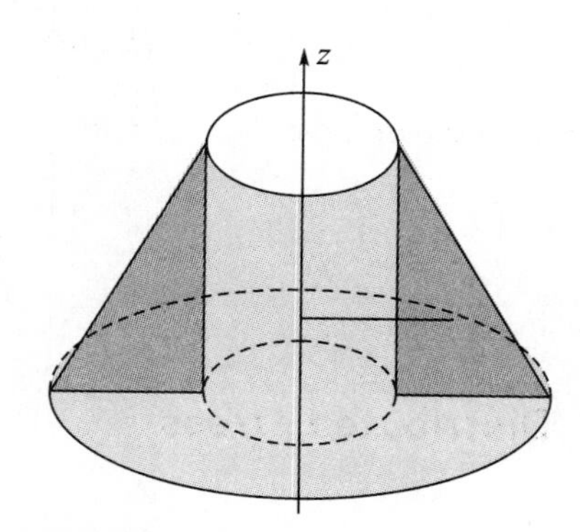

Fig. 8-5 Triangle element

Similar to the three-node element, the equivalent nodal force of the axisymmetrical problem can be expressed as

$$[F]^e = \begin{Bmatrix} F_i^x \\ F_i^y \\ F_j^x \\ F_j^y \\ F_k^x \\ F_k^y \end{Bmatrix} = 2\pi\iint_A \begin{Bmatrix} N_i & 0 \\ 0 & N_i \\ N_j & 0 \\ 0 & N_j \\ N_k & 0 \\ 0 & N_k \end{Bmatrix} \begin{Bmatrix} 0 \\ -\rho g \end{Bmatrix} r\mathrm{d}r\mathrm{d}z = 2\pi\iint_A \begin{Bmatrix} 0 \\ -N_i\rho g \\ 0 \\ -N_j\rho g \\ 0 \\ -N_k\rho g \end{Bmatrix} r\mathrm{d}r\mathrm{d}z \quad (8-21)$$

Eq. (8−21) can be rewritten as

$$\begin{cases} F_i^r = 0 \\ F_i^z = 2\pi\iint_A -N_i\rho g r\mathrm{d}r\mathrm{d}z \end{cases} \quad (i,j,k) \quad (8-22)$$

where $N_i = \frac{1}{2A}(a_i + b_i r + c_i z) \quad (i,j,k)$.

The relationship between area coordinates and Cartesian coordinates can be expressed

$$r = r_i L_i + r_j L_j + r_k L_k \quad (8-23)$$

where $L_i = \frac{A_i}{A}, L_j = \frac{A_j}{A}$ and $L_k = \frac{A_k}{A}$. The following integral equation will be applied

$$\iint_A N_i r\mathrm{d}r\mathrm{d}z = \iint_A L_i(r_i L_i + r_j L_j + r_k L_k)\mathrm{d}r\mathrm{d}z = \frac{A}{12}(2r_i + r_j + r_k)$$

$$= \frac{A}{12}(3\bar{r} + r_i) \quad (i,j,k) \quad (8-24)$$

Substituting Eq. (8−24) into Eq. (8−22), one can have

$$F_i^z = -2\pi\rho g\iint_A N_i r\mathrm{d}r\mathrm{d}z = -\frac{\pi\rho g A}{6}(3\bar{r} + r_i) \quad (i,j,k) \quad (8-25)$$

In solving the above equation, the following formula has been applied

$$\iint L_i^\alpha L_j^\beta L_k^\gamma \mathrm{d}A = \frac{\alpha!\beta!\gamma!}{(\alpha+\beta+\gamma+2)!}2A \quad (8-26)$$

Therefore, one can have

$$\{F_i\} = \begin{Bmatrix} F_i^r \\ F_i^z \end{Bmatrix} = \begin{Bmatrix} 0 \\ -\frac{1}{6}\pi\rho g A(3\bar{r} + r_i) \end{Bmatrix} \quad (i,j,k) \quad (8-27)$$

8.4.2 Distributed stress

The distributed force for axisymmetrical problem must be on the whole annulus because of the symmetrical requirements. For the distributed force as shown in Fig. 8−6, the load can be expressed as

$$\{p\} = \begin{Bmatrix} -q\sin\alpha \\ -q\cos\alpha \end{Bmatrix} = \begin{Bmatrix} -q\frac{z_i - z_k}{l} \\ -q\frac{r_k - r_i}{l} \end{Bmatrix} \quad (8-28)$$

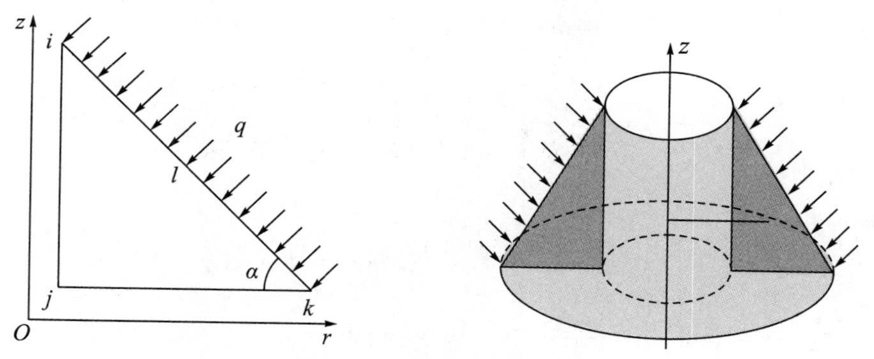

Fig. 8−6 Distributed force

Similarly

$$[F]^e = \begin{Bmatrix} F_i^x \\ F_i^y \\ F_j^x \\ F_j^y \\ F_k^x \\ F_k^y \end{Bmatrix} = 2\pi \int \begin{Bmatrix} N_i & 0 \\ 0 & N_i \\ N_j & 0 \\ 0 & N_j \\ N_k & 0 \\ 0 & N_k \end{Bmatrix} \begin{Bmatrix} -q\dfrac{z_i - z_k}{l} \\ -q\dfrac{r_k - r_i}{l} \end{Bmatrix} r\mathrm{d}s$$

$$= -2\pi\frac{q}{l}\int \begin{Bmatrix} N_i(z_i - z_k) \\ N_i(r_k - r_i) \\ N_j(z_i - z_k) \\ N_j(r_k - r_i) \\ N_k(z_i - z_k) \\ N_k(r_k - r_i) \end{Bmatrix} r\mathrm{d}s = -2\pi\frac{q}{l}\int \begin{Bmatrix} N_i(z_i - z_k) \\ N_i(r_k - r_i) \\ 0 \\ 0 \\ N_k(z_i - z_k) \\ N_k(r_k - r_i) \end{Bmatrix} r\mathrm{d}s \tag{8−29}$$

It should be noted that on the edge ki, as shown in Fig. 8−7, $N_j = L_j = 0$.

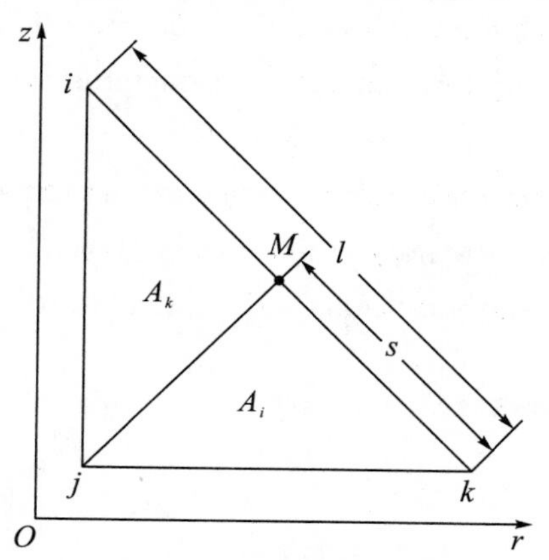

Fig. 8−7 Triangle element

For node i, we have

$$\{F_i\} = \begin{Bmatrix} F_i^r \\ F_i^z \end{Bmatrix} = -2\pi\frac{q}{l}\begin{Bmatrix} (z_i - z_k)\int N_i r\mathrm{d}s \\ (r_k - r_i)\int N_i r\mathrm{d}s \end{Bmatrix} \tag{8−30}$$

For node k, we have

$$\{F_k\}=\begin{Bmatrix}F_k^r\\F_k^z\end{Bmatrix}=-2\pi\frac{q}{l}\begin{Bmatrix}(z_i-z_k)\int N_k r\mathrm{d}s\\(r_k-r_i)\int N_k r\mathrm{d}s\end{Bmatrix} \tag{8-31}$$

According to Eq. (8−23) we have

$$\int N_i r\mathrm{d}s=\int L_i(r_iL_i+r_jL_j+r_kL_k)\mathrm{d}s=\int L_i(r_iL_i+r_kL_k)\mathrm{d}s \tag{8-32}$$

where $L_i=\frac{A_i}{A}=\frac{s}{l}$ and $L_k=\frac{A_k}{A}=\frac{l-s}{l}$. So we have

$$\int r_iL_i^2\mathrm{d}s=\int r_i\left(\frac{s}{l}\right)^2\mathrm{d}s=r_i\frac{l^3}{3l^2}=r_i\frac{l}{3} \tag{8-33}$$

$$\int r_kL_iL_k\mathrm{d}s=\int r_k\left(\frac{l-s}{l}\right)\frac{s}{l}\mathrm{d}s=r_k\int\left(\frac{s}{l}-\frac{s^2}{l^2}\right)\mathrm{d}s=r_k\frac{l}{6} \tag{8-34}$$

Eq. (8−30) can be rewritten as

$$\{F_i\}=\begin{Bmatrix}F_i^r\\F_i^z\end{Bmatrix}=-\frac{\pi q(2r_i+r_k)}{3}\begin{Bmatrix}(z_i-z_k)\\(r_k-r_i)\end{Bmatrix} \tag{8-35}$$

Similarly for node k, we have

$$\{F_k\}=\begin{Bmatrix}F_k^r\\F_k^z\end{Bmatrix}=-2\pi\frac{q}{l}\int\begin{Bmatrix}N_k(z_k-z_i)\\N_k(r_k-r_i)\end{Bmatrix}r\mathrm{d}s=\frac{\pi q}{3}(r_i+2r_k)\begin{Bmatrix}z_i-z_k\\r_k-r_i\end{Bmatrix} \tag{8-36}$$

For node j, because $L_j=0$, so

$$\{F_j\}=\begin{Bmatrix}F_j^r\\F_j^z\end{Bmatrix}=\begin{Bmatrix}0\\0\end{Bmatrix} \tag{8-37}$$

Assignments:

1. When using a triangular cyclic element, what is the analysis procedure?

2. Is the three-node triangular element for isosymmetrical problem a constant strain element? And why?

3. There are two axisymmetric right triangular elements as shown in the following figure. The shape, size and orientation are same, but the location is different. The elasticity modulus is E, Poisson's ratio is 0.15. Calculate their element stiffness matrix, respectively.

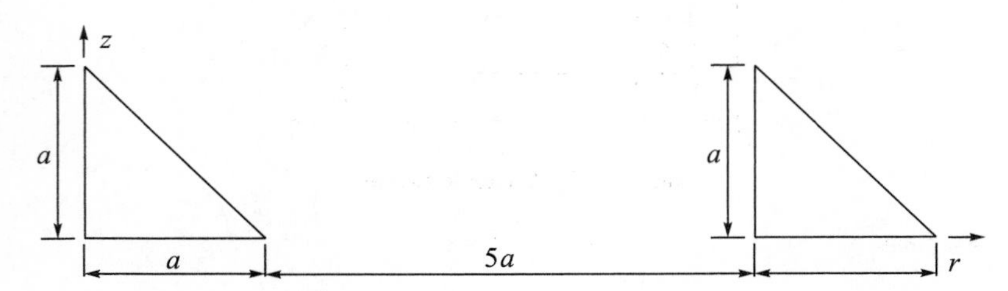

Chapter 9 Three-dimensional stress analysis

In the previous chapters, we have learnt two dimensional (plane problem) finite element method. In the engineering practice, the structures or domain may not be precisely analyzed by using two dimensional finite element method. In this chapter, three dimensional (3D) finite element method will be studied, and we will first learn tetrahedron element method, and then study brick element method.

It is immediately obvious, however, that 3D elements will result in very large numbers of simultaneous equations in practical problems, which may place a severe limitation on the use of the method in practice. Further, the bandwidth of the resulting equation system becomes large, leading to increased use of iterative solution methods.

9.1 Tetrahedron element method

The simplest two-dimensional continuum element is the three-node triangle. In three dimensions its equivalent is a tetrahedron, an element with four nodal corners, and this chapter will deal with the basic formulation of such an element as shown in Fig. 9—1.

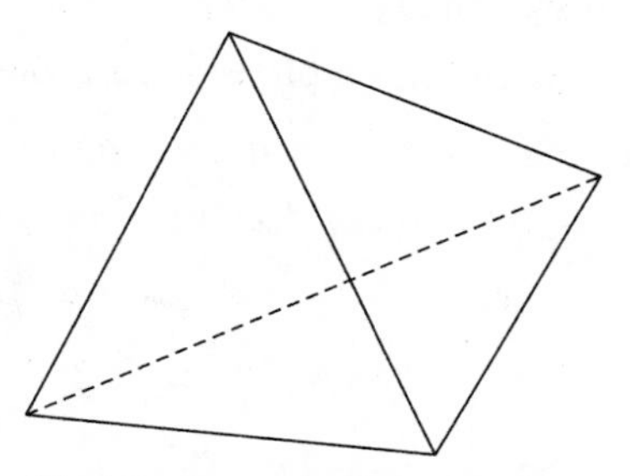

Fig. 9—1 A tetrahedral element

Immediately, a difficulty not encountered previously is presented. It is one of ordering of the nodal numbers and, in fact, of a suitable representation of a body divided into such elements. Fig. 9—2 shows a tetrahedral element numbered with i, j, m, p. We should always use a consistent order of numbering, e. g., for p count the other nodes in an anticlockwise order as viewed from p, giving the element as i, j, m, p, i. e. the right-hand rule.

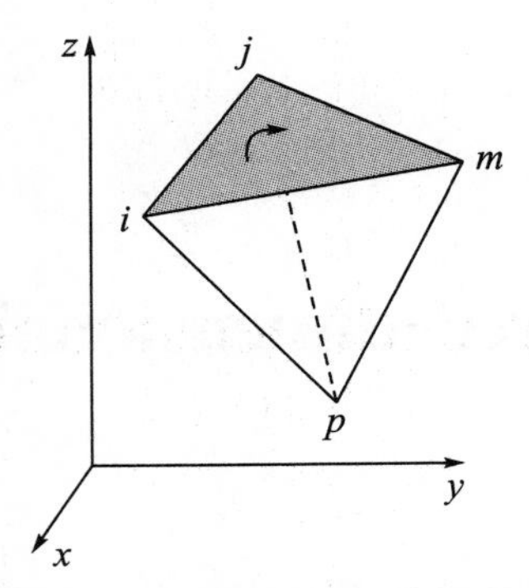

Fig. 9-2 A tetrahedral element

Each node has three displacement components, i. e.

$$\begin{Bmatrix} u \\ v \\ w \end{Bmatrix} \tag{9-1}$$

Therefore, each element has 12 displacement components, i. e.

$$\{\delta\}^e = \{u_i \quad v_i \quad w_i \quad u_j \quad v_j \quad w_j \quad u_m \quad v_m \quad w_m \quad u_p \quad v_p \quad w_p\}^{\mathrm{T}} \tag{9-2}$$

The displacement function can be written as

$$\begin{cases} u = \alpha_1 + \alpha_2 x + \alpha_3 y + \alpha_4 z \\ v = \alpha_5 + \alpha_6 x + \alpha_7 y + \alpha_8 z \\ w = \alpha_9 + \alpha_{10} x + \alpha_{11} y + \alpha_{12} z \end{cases} \tag{9-3}$$

The 12 unknowns can be determined from the four node displacements. For the x-direction displacements, we have

$$\begin{cases} u_i = \alpha_1 + \alpha_2 x_i + \alpha_3 y_i + \alpha_4 z_i \\ u_j = \alpha_1 + \alpha_2 x_j + \alpha_3 y_j + \alpha_4 z_j \\ u_m = \alpha_1 + \alpha_2 x_m + \alpha_3 y_m + \alpha_4 z_m \\ u_p = \alpha_1 + \alpha_2 x_p + \alpha_3 y_p + \alpha_4 z_p \end{cases} \rightarrow \begin{Bmatrix} u_i \\ u_j \\ u_m \\ u_p \end{Bmatrix} = \begin{bmatrix} 1 & x_i & y_i & z_i \\ 1 & x_j & y_j & z_j \\ 1 & x_m & y_m & z_m \\ 1 & x_p & y_p & z_p \end{bmatrix} \begin{Bmatrix} \alpha_1 \\ \alpha_2 \\ \alpha_3 \\ \alpha_4 \end{Bmatrix} \tag{9-4}$$

Similarly, for y-direction and z-direction displacements, one can have

$$\begin{cases} v_i = \alpha_5 + \alpha_6 x_i + \alpha_7 y_i + \alpha_8 z_i \\ v_j = \alpha_5 + \alpha_6 x_j + \alpha_7 y_j + \alpha_8 z_j \\ v_m = \alpha_5 + \alpha_6 x_m + \alpha_7 y_m + \alpha_8 z_m \\ v_p = \alpha_5 + \alpha_6 x_p + \alpha_7 y_p + \alpha_8 z_p \end{cases} \tag{9-5}$$

and

$$\begin{cases} w_i = \alpha_9 + \alpha_{10} x_i + \alpha_{11} y_i + \alpha_{12} z_i \\ w_j = \alpha_9 + \alpha_{10} x_j + \alpha_{11} y_j + \alpha_{12} z_j \\ w_m = \alpha_9 + \alpha_{10} x_m + \alpha_{11} y_m + \alpha_{12} z_m \\ w_p = \alpha_9 + \alpha_{10} x_p + \alpha_{11} y_p + \alpha_{12} z_p \end{cases} \tag{9-6}$$

After solving the coefficients α, substitute α into the displacement function, and one can have

$$\begin{cases} u = N_i u_i + N_j u_j + N_m u_m + N_p u_p \\ v = N_i v_i + N_j v_j + N_m v_m + N_p v_p \\ w = N_i w_i + N_j w_j + N_m w_m + N_p w_p \end{cases} \tag{9-7}$$

where

$$N_i = \frac{a_i + b_i x + c_i y + d_i z}{6V} \tag{9-8}$$

and

$$6V = \det\begin{vmatrix} 1 & x_i & y_i & z_i \\ 1 & x_j & y_j & z_j \\ 1 & x_m & y_m & z_m \\ 1 & x_p & y_p & z_p \end{vmatrix}$$

$$a_i = \det\begin{vmatrix} x_j & y_j & z_j \\ x_m & y_m & z_m \\ x_p & y_p & z_p \end{vmatrix} \qquad b_i = -\det\begin{vmatrix} 1 & y_j & z_j \\ 1 & y_m & z_m \\ 1 & y_p & z_p \end{vmatrix}$$

$$c_i = -\det\begin{vmatrix} x_j & 1 & z_j \\ x_m & 1 & z_m \\ x_p & 1 & z_p \end{vmatrix} \qquad d_i = -\det\begin{vmatrix} x_j & y_j & 1 \\ x_m & y_m & 1 \\ x_p & y_p & 1 \end{vmatrix}$$

The displacements can be rewritten as

$$\begin{Bmatrix} u \\ v \\ w \end{Bmatrix} = [N]\{\delta\}^e = \begin{bmatrix} N_i & 0 & 0 & N_j & 0 & 0 & N_m & 0 & 0 & N_p & 0 & 0 \\ 0 & N_i & 0 & 0 & N_j & 0 & 0 & N_m & 0 & 0 & N_p & 0 \\ 0 & 0 & N_i & 0 & 0 & N_j & 0 & 0 & N_m & 0 & 0 & N_p \end{bmatrix} \begin{Bmatrix} u_i \\ v_i \\ w_i \\ u_j \\ v_j \\ w_j \\ u_m \\ v_m \\ w_m \\ u_p \\ v_p \\ w_p \end{Bmatrix} \tag{9-9}$$

where the shape function N_i, N_j, N_m and N_p can be obtained from Eq. (9−9).

9.1.1 Strain Matrix

The relationship of trains versus displacements can be expressed as

$$\{\varepsilon\} = \begin{Bmatrix} \varepsilon_x \\ \varepsilon_y \\ \varepsilon_z \\ \gamma_{xy} \\ \gamma_{yz} \\ \gamma_{zx} \end{Bmatrix} = \begin{Bmatrix} \frac{\partial u}{\partial x} \\ \frac{\partial v}{\partial y} \\ \frac{\partial w}{\partial z} \\ \frac{\partial u}{\partial y} + \frac{\partial v}{\partial x} \\ \frac{\partial v}{\partial z} + \frac{\partial w}{\partial y} \\ \frac{\partial w}{\partial x} + \frac{\partial u}{\partial z} \end{Bmatrix} \tag{9-10}$$

Substituting the displacements in Eq. (9—9) into Eq. (9—10), one can get

$$\{\varepsilon\} = [B]\{\delta\}^e = [B_i \quad B_j \quad B_m \quad B_p]\{\delta\}^e \tag{9-11}$$

where

$$[B] = \frac{1}{6V}\begin{bmatrix} b_i & 0 & 0 & b_j & 0 & 0 & b_m & 0 & 0 & b_p & 0 & 0 \\ 0 & c_i & 0 & 0 & c_j & 0 & 0 & c_m & 0 & 0 & c_p & 0 \\ 0 & 0 & d_i & 0 & 0 & d_j & 0 & 0 & d_m & 0 & 0 & d_p \\ c_i & b_i & 0 & c_j & b_j & 0 & c_m & b_m & 0 & c_p & b_p & 0 \\ 0 & d_i & c_i & 0 & d_j & c_j & 0 & d_m & c_m & 0 & d_p & c_p \\ d_i & 0 & b_i & d_j & 0 & b_j & d_m & 0 & b_m & d_p & 0 & b_p \end{bmatrix} \tag{9-12}$$

9. 1. 2 Stress Matrix

For three dimensional problem, the $[D]$ matrix is

$$[D] = \frac{E}{(1+\nu)(1-2\nu)}\begin{bmatrix} 1-\nu & \nu & \nu & 0 & 0 & 0 \\ & 1-\nu & \nu & 0 & 0 & 0 \\ & & 1-\nu & 0 & 0 & 0 \\ & & & \frac{1-2\nu}{2} & 0 & 0 \\ & Sym & & & \frac{1-2\nu}{2} & 0 \\ & & & & & \frac{1-2\nu}{2} \end{bmatrix} \tag{9-13}$$

The stress matrix can be written as

$$\{\sigma\} = \begin{Bmatrix} \sigma_x \\ \sigma_y \\ \sigma_z \\ \tau_{xy} \\ \tau_{yz} \\ \tau_{zx} \end{Bmatrix} = [D]\{\varepsilon\}^e = [D][B]\{\delta\}^e = [S]\{\delta\}^e \tag{9-14}$$

Substituting the $[D]$ matrix in Eq. (9−13) and the $[B]$ matrix in Eq. (9−12) into Eq. (9−14), one can get the $[S]$ matrix. Because in the $[B]$ matrix, all the coefficients a, b, c and d are constants, $[B]$ is a constant matrix, and the $[S]$ is also a constant matrix, which is caused by the linear displacement functions selected in Eq. (9−3).

9.1.3 Element Stiffness Matrix

According to the principle of virtual displacement, one can have

$$(\{\delta^*\}^e)^T\{F\}^e = \iiint \{\varepsilon^*\}^T\{\sigma\}\mathrm{d}x\mathrm{d}y\mathrm{d}z \tag{9-15}$$

where $\{\varepsilon^*\} = [B]\{\delta^*\}^e$ and $\{\sigma\} = [D][B]\{\delta\}^e$. Substituting them into Eq. (9−15), one can have

$$(\{\delta^*\}^e)^T\{F\}^e = \iiint (\{\delta^*\}^e)^T[B]^T[D][B]\{\delta\}^e\mathrm{d}x\mathrm{d}y\mathrm{d}z \tag{9-16}$$

Deleting the nodal virtual displacements $(\{\delta^*\}^e)^T$, Eq. (9−16) can be rewritten as

$$\{F\}^e = \iiint [B]^T[D][B]\mathrm{d}x\mathrm{d}y\mathrm{d}z \cdot \{\delta\}^e = [K]^e\{\delta\}^e \tag{9-17}$$

where $[K]^e$ is element stiffness matrix which can be written as

$$[K]^e = \iiint [B]^T[D][B]\mathrm{d}x\mathrm{d}y\mathrm{d}z = [B]^T[D][B]V \tag{9-18}$$

Substituting $[B]$ and $[D]$ into Eq. (9−18), one can have

$$[K]^e = \begin{bmatrix} K_{ii} & -K_{ij} & K_{im} & -K_{ip} \\ -K_{ji} & K_{jj} & -K_{jm} & K_{jp} \\ K_{mi} & -K_{mj} & K_{mm} & -K_{mp} \\ -K_{pi} & K_{pj} & -K_{pm} & K_{pp} \end{bmatrix} \tag{9-19}$$

where

$$[K_{rs}] = \frac{E(1-\mu)}{36(1+\mu)(1-2\mu)V}\begin{bmatrix} b_rb_s+A_2(c_rc_s+d_rd_s) & A_1b_rc_s+A_2c_rb_s & A_1b_rd_s+A_2d_rb_s \\ A_1c_rb_s+A_2b_rc_s & c_rc_s+A_2(b_rb_s+d_rd_s) & A_1c_rd_s+A_2d_rc_s \\ A_1d_rb_s+A_2b_rd_s & A_1d_rc_s+A_2c_rd_s & d_rd_s+A_2(b_rb_s+c_rc_s) \end{bmatrix} \tag{9-20}$$

where $A_1 = \mu/(1-\mu)$, $A_2 = (1-2\mu)/2(1-\mu)$, $a_i = \det\begin{vmatrix} x_j & y_j & z_j \\ x_m & y_m & z_m \\ x_p & y_p & z_p \end{vmatrix}$,

$b_i = -\det\begin{vmatrix} 1 & y_j & z_j \\ 1 & y_m & z_m \\ 1 & y_p & z_p \end{vmatrix}$, $c_i = -\det\begin{vmatrix} x_j & 1 & z_j \\ x_m & 1 & z_m \\ x_p & 1 & z_p \end{vmatrix}$, $d_i = -\det\begin{vmatrix} x_j & y_j & 1 \\ x_m & y_m & 1 \\ x_p & y_p & 1 \end{vmatrix}$

and $6V = \det\begin{vmatrix} 1 & x_i & y_i & z_i \\ 1 & x_j & y_j & z_j \\ 1 & x_m & y_m & z_m \\ 1 & x_p & y_p & z_p \end{vmatrix}$.

We can find that all the terms in the element stiffness matrix can be determined, and $[K]^e$ is a constant matrix.

9.1.4 Global stiffness matrix

Fig. 9 − 3 shows three tetrahedron elements, and the relationship between the element stiffness and the global stiffness is illustrated in the following.

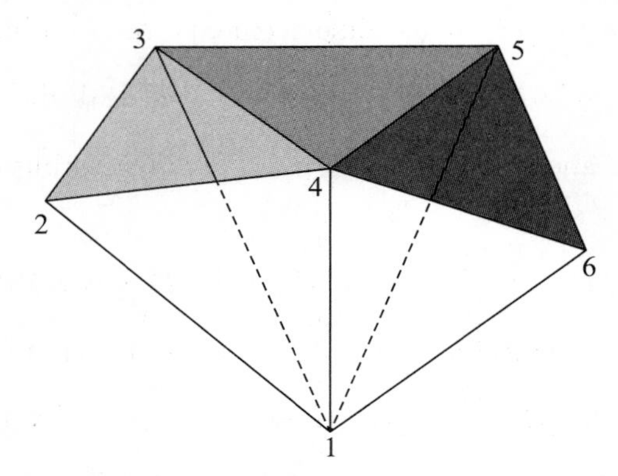

Fig. 9−3 Three tetrahedron element

For a single element, the element stiffness matrix is shown in Eq. (9−19). Similar to the three-node triangle element, after element stiffness matrix is obtained, we install all the terms into the global stiffness matrix according to the real numbers, and finally the global equilibrium equation can be written as

$$\begin{bmatrix} K_{11} & K_{12} & K_{13} & K_{14} & K_{15} & K_{16} \\ K_{21} & K_{22} & K_{23} & K_{24} & & \\ K_{31} & K_{32} & K_{33} & K_{34} & K_{35} & \\ K_{41} & K_{42} & K_{43} & K_{44} & K_{45} & K_{46} \\ K_{51} & & K_{53} & K_{54} & K_{55} & K_{56} \\ K_{61} & & & K_{64} & K_{65} & K_{66} \end{bmatrix} \begin{Bmatrix} \{\delta_1\} \\ \{\delta_2\} \\ \{\delta_3\} \\ \{\delta_4\} \\ \{\delta_5\} \\ \{\delta_6\} \end{Bmatrix} = \begin{Bmatrix} \{F_1\} \\ \{F_2\} \\ \{F_3\} \\ \{F_4\} \\ \{F_5\} \\ \{F_6\} \end{Bmatrix} \tag{9−21}$$

9.2 Volume coordinates

For two dimensional triangle element, we have learnt area coordinates. For 3D tetrahedron element, we will learn volume coordinates. Fig. 9−4 shows a point inside the tetrahedral element, and this point connecting with the four nodes forming four tetrahedrons $P234$, $P341$, $P412$ and $P123$. We use V_1, V_2, V_3 and V_4 denote the volumes of the tetrahedron $P234$, $P341$, $P412$ and $P123$, respectively.

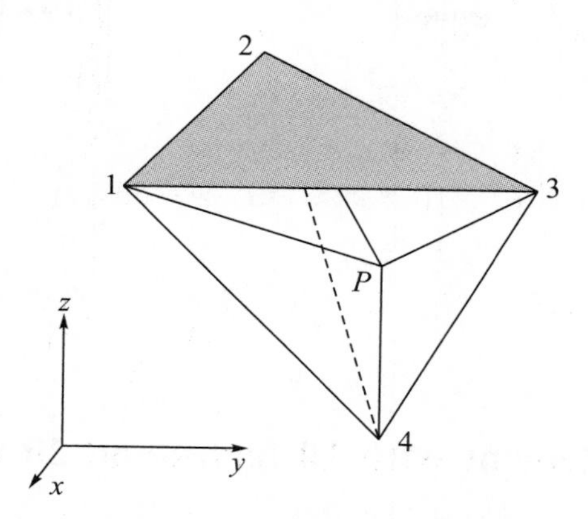

Fig. 9—4 Tetrahedron element

The volume coordinates are defined as

$$L_1 = \frac{V_1}{V},\ L_2 = \frac{V_2}{V},\ L_3 = \frac{V_3}{V},\ L_4 = \frac{V_4}{V} \tag{9-22}$$

where V is the volume of the tetrahedron element, and V can be obtained by

$$6V = \det\begin{vmatrix} 1 & x_i & y_i & z_i \\ 1 & x_j & y_j & z_j \\ 1 & x_m & y_m & z_m \\ 1 & x_p & y_p & z_p \end{vmatrix}$$

The volume coordinates have the property

$$L_1 + L_2 + L_3 + L_4 = 1 \tag{9-23}$$

It is easy proved because

$$V_1 + V_2 + V_3 + V_4 = V \tag{9-24}$$

The relationship of volume coordinates vs. the Cartesian coordinates can be written as

$$\begin{Bmatrix} 1 \\ x \\ y \\ z \end{Bmatrix} = \begin{bmatrix} 1 & 1 & 1 & 1 \\ x_1 & x_2 & x_3 & x_4 \\ y_1 & y_2 & y_3 & y_4 \\ z_1 & z_2 & z_3 & z_4 \end{bmatrix} \begin{Bmatrix} L_1 \\ L_2 \\ L_3 \\ L_4 \end{Bmatrix} \tag{9-25}$$

From Eq. (9—25), we have

$$\begin{Bmatrix} L_1 \\ L_2 \\ L_3 \\ L_4 \end{Bmatrix} = \frac{1}{6V} \begin{bmatrix} a_1 & b_1 & c_1 & d_1 \\ a_2 & b_2 & c_2 & d_2 \\ a_3 & b_3 & c_3 & d_3 \\ a_4 & b_4 & c_4 & d_4 \end{bmatrix} \begin{Bmatrix} 1 \\ x \\ y \\ z \end{Bmatrix} \tag{9-26}$$

The coefficients of a, b, c and d can be obtained from the following formulae

$$a_i = \det\begin{vmatrix} x_j & y_j & z_j \\ x_m & y_m & z_m \\ x_p & y_p & z_p \end{vmatrix} \qquad b_i = -\det\begin{vmatrix} 1 & y_j & z_j \\ 1 & y_m & z_m \\ 1 & y_p & z_p \end{vmatrix}$$

$$c_i = -\det\begin{vmatrix} x_j & 1 & z_j \\ x_m & 1 & z_m \\ x_p & 1 & z_p \end{vmatrix} \qquad d_i = -\det\begin{vmatrix} x_j & y_j & 1 \\ x_m & y_m & 1 \\ x_p & y_p & 1 \end{vmatrix}$$

Apparently, $L_1 = N_1$, $L_2 = N_2$, $L_3 = N_3$, $L_4 = N_4$.

9.3 Tetrahedral element with 10 nodes and 20 nodes

Tetrahedron element with 10 nodes and 20 nodes will be studied.

9.3.1 Tetrahedron element with 10 nodes

The tetrahedral element with 10 nodes is shown as Fig. 9—5.

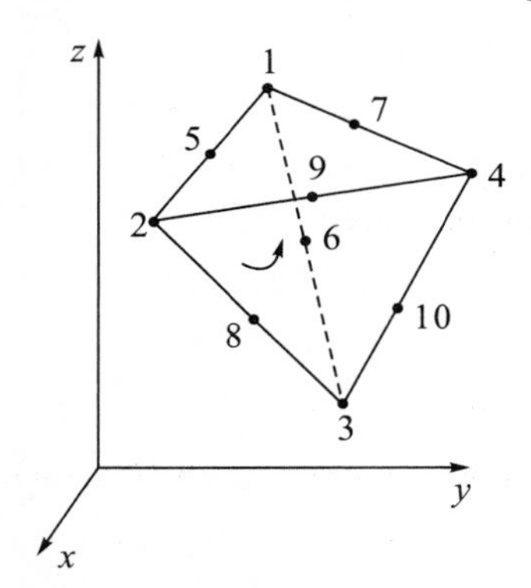

Fig. 9—5 Ten-node tetrahedron element

There are 10 nodes, and each node has 3 displacements, and therefore, there are totally 30 equations which can be used, that means the displacement functions can have the maximum 30 coefficients. According to the Pascal triangle rule, the displacement functions of the ten-node tetrahedron element can be expressed as

$$\begin{cases} u = \alpha_1 + \alpha_2 x + \alpha_3 y + \alpha_4 z + \alpha_5 x^2 + \alpha_6 y^2 + \alpha_7 z^2 + \alpha_8 xy + \alpha_9 yz + \alpha_{10} zx \\ v = \alpha_{11} + \alpha_{12} x + \alpha_{13} y + \alpha_{14} z + \alpha_{15} x^2 + \alpha_{16} y^2 + \alpha_{17} z^2 + \alpha_{18} xy + \alpha_{19} yz + \alpha_{20} zx \\ w = \alpha_{21} + \alpha_{22} x + \alpha_{23} y + \alpha_{24} z + \alpha_{25} x^2 + \alpha_{26} y^2 + \alpha_{27} z^2 + \alpha_{28} xy + \alpha_{29} yz + \alpha_{30} zx \end{cases} \tag{9—27}$$

After the 30 coefficients have been solved, substitute them into Eq. (9—27), and the displacement function can be expressed as

$$u = \sum_{i=1}^{10} N_i u_i, \quad v = \sum_{i=1}^{10} N_i v_i, \quad w = \sum_{i=1}^{10} N_i w_i \tag{9—28}$$

where

$$\begin{cases} N_i = (2L_i - 1)L_i \quad (i = 1,2,3,4) \\ N_j = 4L_mL_p \quad (j \text{ is situated in edge center}) \end{cases} \tag{9-29}$$

and

$$L_1 = \frac{V_1}{V}, \ L_2 = \frac{V_2}{V}, \ L_3 = \frac{V_3}{V}, \ L_4 = \frac{V_4}{V} \tag{9-30}$$

Similar as the four-node tetrahedron element, the strain matrix, stress matrix, element stiffness matrix and the global stiffness can be obtained.

9.3.2 Tetrahedron element with 20 nodes

We design 2 nodes in each edge and 1 node in each triangle center, thus there are totally 20 nodes as shown in Fig. 9-6.

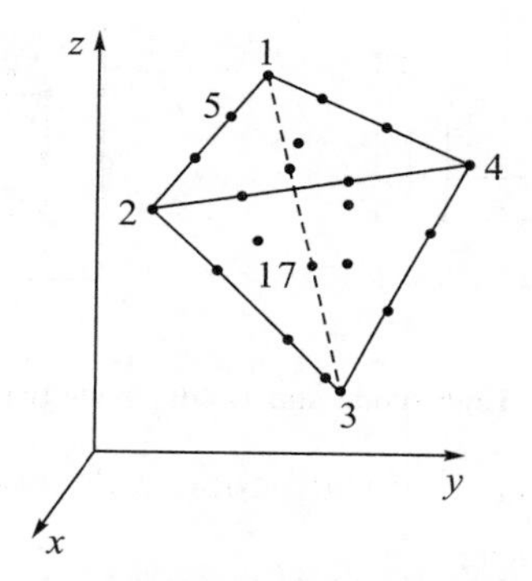

Fig. 9-6 Tetrahedron element with 20 nodes

For each node, there are 3 displacements, so totally 60 equations can be established. The displacement functions in x-axis direction can be written as

$$u = \alpha_1 + \alpha_2 x + \alpha_3 y + \alpha_4 z + \alpha_5 x^2 + \alpha_6 y^2 + \alpha_7 z^2 + \alpha_8 xy + \alpha_9 yz + \alpha_{10} zx + \alpha_{11} x^3 + \alpha_{12} y^3 + \alpha_{13} z^3 + \alpha_{14} x^2 y + \alpha_{15} x^2 z + \alpha_{16} y^2 x + \alpha_{17} y^2 z + \alpha_{18} z^2 x + \alpha_{19} z^2 y + \alpha_{20} xyz \tag{9-31}$$

The corresponding shape functions can be written as

$$u = \sum_{i=1}^{20} N_i u_i, \ v = \sum_{i=1}^{20} N_i v_i, \ w = \sum_{i=1}^{20} N_i w_i \tag{9-32}$$

where for the nodes

$$N_1 = \frac{1}{2}(3L_1 - 1)(3L_1 - 2)L_1 \quad (1,\ 2,\ 3,\ 4)$$

For the edges

$$N_5 = \frac{9}{2}L_1L_2(3L_1 - 1) \quad (5,\ 6,\ \cdots,16)$$

For the triangle center

$$N_{17} = 27L_1L_2L_3 \quad (17,\ 18,\ 19,\ 20)$$

Similar to the four-node tetrahedron element, the strain matrix, stress matrix, element stiffness matrix and the global stiffness can be obtained.

9.4 Brick element

The division of space volume into individual tetrahedra sometimes presents difficulties of visualization and could easily lead to errors in nodal numbering, etc., unless a fully automatic code is available. A more convenient subdivision of space is into eight-cornered brick elements (brick being the natural way to building a universe).

The 8 node element and 20 node elementare shown in Fig. 9—7, and they are widely used elements in three dimensional stress analysis.

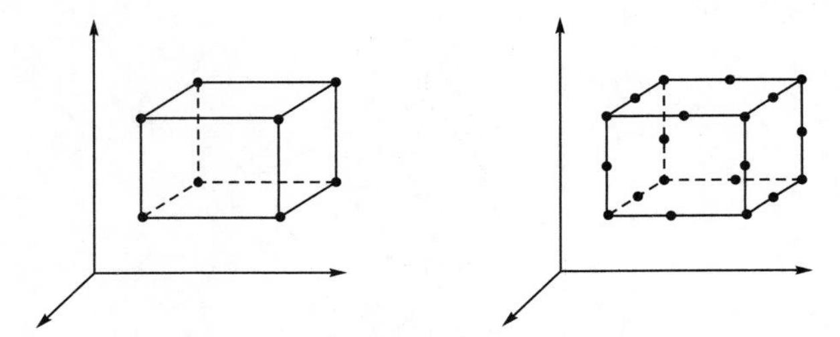

Fig. 9—7 Eight-node and twenty-node brick elements

For the eight-node element, there are totally 24 nodal displacements can be used, and the displacement functions are

$$\begin{cases} u = \alpha_1 + \alpha_2\xi + \alpha_3\eta + \alpha_4\zeta + \alpha_5\xi\eta + \alpha_6\eta\xi + \alpha_7\xi\zeta + \alpha_8\xi\eta\zeta \\ v = \alpha_9 + \alpha_{10}\xi + \alpha_{11}\eta + \alpha_{12}\zeta + \alpha_{13}\xi\eta + \alpha_{14}\eta\xi + \alpha_{15}\xi\zeta + \alpha_{16}\xi\eta\zeta \\ w = \alpha_{17} + \alpha_{18}\xi + \alpha_{19}\eta + \alpha_{20}\zeta + \alpha_{21}\xi\eta + \alpha_{22}\eta\xi + \alpha_{23}\xi\zeta + \alpha_{24}\xi\eta\zeta \end{cases} \tag{9—33}$$

The displacements can be simplified as

$$\begin{cases} u = \sum_{i=1}^{8} N_i(\xi, \eta, \zeta)u_i \\ v = \sum_{i=1}^{8} N_i(\xi, \eta, \zeta)v_i \\ w = \sum_{i=1}^{8} N_i(\xi, \eta, \zeta)w_i \end{cases} \tag{9—34}$$

The corresponding shape functions can be expressed as

$$N_i(\xi, \eta, \zeta) = \frac{1}{8}(1+\xi_i\xi)(1+\eta_i\eta)(1+\zeta_i\zeta) \quad (i = 1, 2, \cdots, 8) \tag{9—35}$$

For the twenty-node brick element, there are totally 60 nodal displacements can be used, and the displacements in x-axis direction can be expressed as

$$u = \alpha_1 + \alpha_2\xi + \alpha_3\eta + \alpha_4\zeta + \alpha_5\xi^2 + \alpha_6\eta^2 + \alpha_7\zeta^2 + \alpha_8\xi\eta + \alpha_9\eta\zeta + \alpha_{10}\xi\zeta +$$
$$\alpha_{11}\xi^2\eta + \alpha_{12}\xi^2\zeta + \alpha_{13}\eta^2\xi + \alpha_{14}\eta^2\zeta + \alpha_{15}\zeta^2\xi + \alpha_{16}\zeta^2\eta + \alpha_{17}\xi\eta\zeta +$$

$$\alpha_{18}\xi^2\eta\zeta + \alpha_{19}\xi\eta^2\zeta + \alpha_{20}\xi\eta\zeta^2 \tag{9-36}$$

The simplified expression is

$$\begin{cases} u = \sum_{i=1}^{20} N_i(\xi, \eta, \zeta)u_i \\ v = \sum_{i=1}^{20} N_i(\xi, \eta, \zeta)v_i \\ w = \sum_{i=1}^{20} N_i(\xi, \eta, \zeta)w_i \end{cases} \tag{9-37}$$

The shape functions are

$$\begin{cases} N_i = \frac{1}{8}(1+\xi_i\xi)(1+\eta_i\eta)(1+\zeta_i\zeta)(\xi_i\xi+\eta_i\eta+\zeta_i\zeta-2) \\ (i = 1, 3, 5, 7, 13, 15, 17, 19) \\ N_i = \frac{1}{4}(1-\xi^2)(1+\eta_i\eta)(1+\zeta_i\zeta) \quad (i = 2, 6, 14, 18) \\ N_i = \frac{1}{4}(1-\eta^2)(1+\xi_i\xi)(1+\zeta_i\zeta) \quad (i = 4, 8, 16, 20) \\ N_i = \frac{1}{4}(1-\zeta^2)(1+\eta_i\eta)(1+\xi_i\xi) \quad (i = 9, 10, 11, 12) \end{cases} \tag{9-38}$$

From the above displacement functions, we find that the strain matrix and stress matrix must contain the variables ξ, η and ζ, and the integrand of the element stiffness cannot move out. Therefore, in the next chapter, we will learn Gauss integration method which can deal with such problems.

Assignments:

1. Eight-node brick element is shown in the following figure. Each edge length is 2 unit, apply uniform load q in the vertical direction on the surface of $\zeta = 1$. Calculate the equivalent node loads.

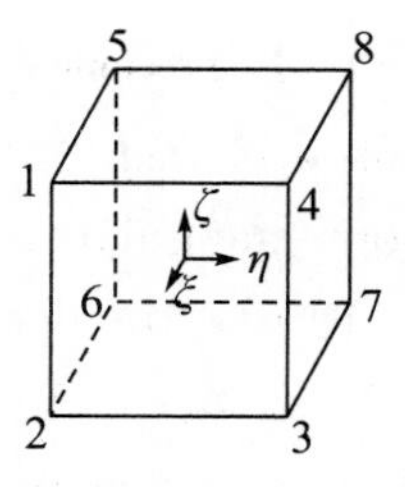

2. State the advantage and disadvantage of tetrahedron element in finite element method.

Chapter 10 Isoparametric element

10.1 Definition of isoparametric element

In this chapter, we will study isoparametric element, which may appear somewhat tedious (and confusing initially), but it will lead to a simple computer program formulation, and it generally applicable for two and three-dimensional stress analysis and for nonstructural problems. The isoparametric formulation allows elements to be created that are nonrectangular and have curved sides, which will be difficult to use the former displacement functions to study.

The concept of isoparametric is derived from the use of the same shape functions (parameters) to define the element shape (or position) as used to define the displacement within the element, i. e. both the coordinate transformations and displacements use the same parameters.

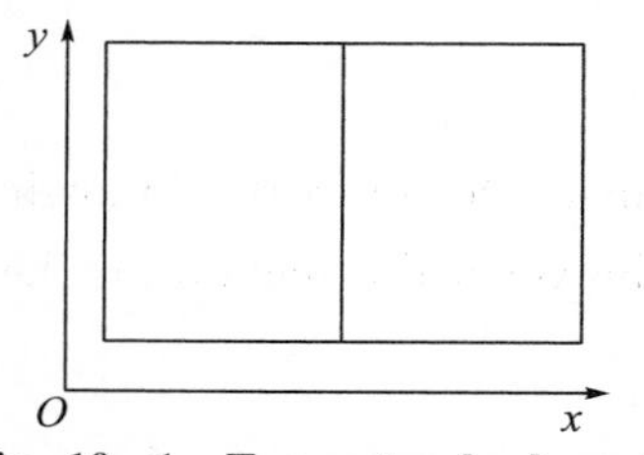

Fig. 10−1 Two rectangle elements

Why do we study the isoparametric element? The isoparametric formulation allows elements to be created that are nonrectangular and have curved sides, e. g. for a rectangular element as shown in Fig. 10−1, the displacement functions are

$$\begin{cases} u = \alpha_1 + \alpha_2 x + \alpha_3 y + \alpha_4 xy \\ v = \alpha_5 + \alpha_6 x + \alpha_7 y + \alpha_8 xy \end{cases} \tag{10-1}$$

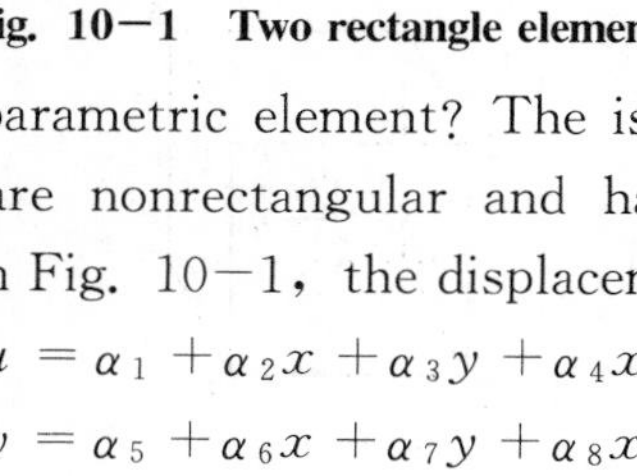

Fig. 10−2 Two quadrilateral elements

If we use them to describe an arbitrary shaped quadrilateral, as showing in Fig. 10−2, for edge AB, we can have $y=kx+b$; then the displacement function can be written as

$$u = Ax^2 + Bx + C \tag{10−2}$$

It can be seen that the displacement is not linear anymore, and therefore, the displacement along the mutual edge AB cannot be determined only by the two nodes A and B, that may lead to two cases of dislocation, overlap and gap. The displacements compatibility along AB cannot be guaranteed, that indicates the above displacement functions may cause dislocation occurring along the edge AB. However, if the edge AB is parallel to x-axis or y-axis, then $y=C$ or $x=C$, and the displacements in Eq. (10−1) u and v are actually linearly related to x and y, that can guarantee the displacement continuity on the mutual edge. This chapter is therefore concerned with the subject of distorting elements or more arbitrary shaped elements.

10.2 Mapping method

In order to solve the distorting element problem, mapping technique will be applied here. A 2D square element will be 'mapped' into distorted form in the manner indicated in Fig. 10−3. It is shown that the ξ, η coordinates can be distorted to a new curvilinear set when plotted in Cartesian x, y space.

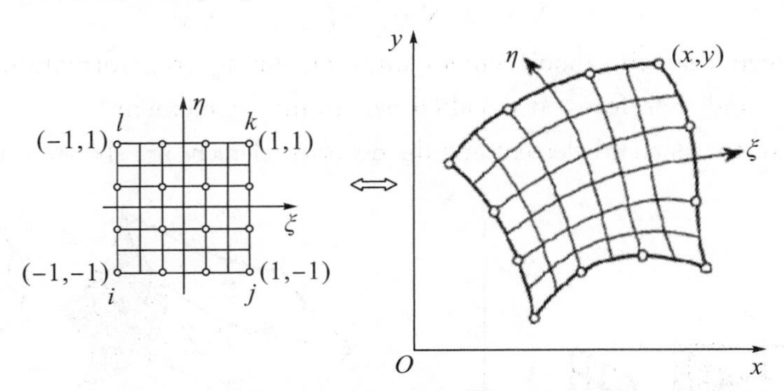

Fig. 10−3 A 2D element is mapped into a distorted element

The purpose is to find the relationship between the square element and the distorted element. The studying on the distorted element will be therefore changed to the studying on the square element.

A most convenient method of establishing the coordinate transformations is to use the shape functions we have already derived to represent the variation of the unknown function.

If we write the x, y coordinates, for instance, for each element as

$$\begin{cases} x = N_i x_i + N_j x_j + N_k x_k + N_l x_l \\ y = N_i y_i + N_j y_j + N_k y_k + N_l y_l \end{cases} \tag{10−3}$$

where N_i, N_j, N_k and N_l are the shape functions of the square element, which can be written as

$$\begin{cases} N_i = \frac{1}{4}(1-\xi)(1-\eta) \\ N_j = \frac{1}{4}(1+\xi)(1-\eta) \\ N_k = \frac{1}{4}(1+\xi)(1+\eta) \\ N_l = \frac{1}{4}(1-\xi)(1+\eta) \end{cases} \tag{10-4}$$

At point i $(-1, -1)$, $N_i=1$, $N_j=N_k=N_l=0$, and then from Eq. (10−3), we have $x=x_i$, $y=y_i$. At the center point (0, 0), $N_i=N_j=N_k=N_l=1/4$, and from Eq. (10.3), $x=(x_i+x_j+x_k+x_l)/4$, $y=(y_i+y_j+y_k+y_l)/4$, which is just the center of the distorted element.

To each set of local coordinates (ξ, η), there will correspond to a set of global Cartesian coordinates (x, y) and in general only one such set. We shall see, however, that a non-uniqueness may arise sometimes with violent distortion.

The concept of using such element shape functions for establishing curvilinear coordinates in the context of finite element analysis appears to have been first introduced by Taig. In his first application basic linear quadrilateral relations were used. Irons generalized the idea for other elements.

The displacements for the distorted element can be written as

$$\begin{cases} u = N_i u_i + N_j u_j + N_k u_k + N_l u_l \\ v = N_i v_i + N_j v_j + N_k v_k + N_l v_l \end{cases} \tag{10-5}$$

It can be seen that the displacement and coordinate transformations use the same shape functions, and therefore, it is called isoparametric element.

For 3D cases, the standard element and the distorted element are shown in Fig. 10−4.

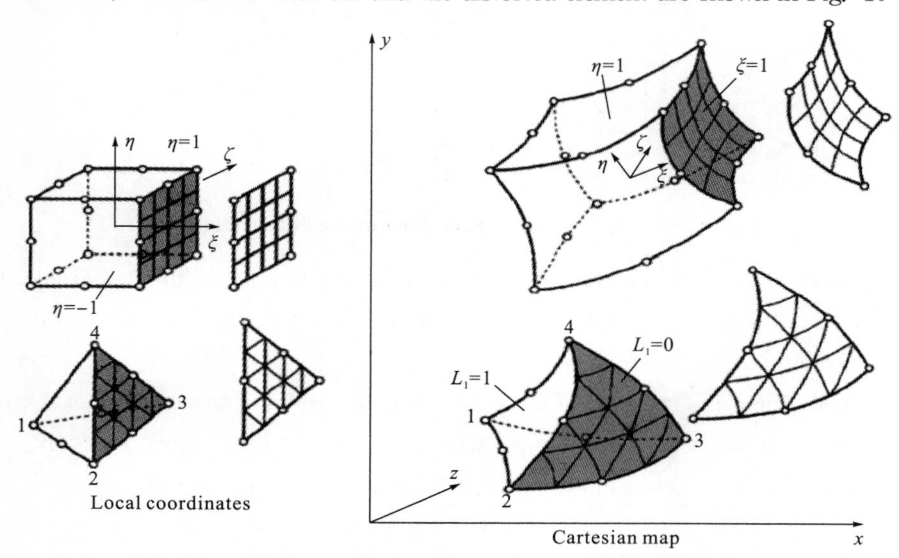

Fig. 10−4 Mapping of standard element into distorted element

For triangle element, we have learnt the area coordinates, and

$$\begin{cases} x_iL_i = \dfrac{1}{2A}(a_i + b_ix + c_iy)x_i \\ x_jL_j = \dfrac{1}{2A}(a_j + b_jx + c_jy)x_j \\ x_kL_k = \dfrac{1}{2A}(a_k + b_kx + c_ky)x_k \end{cases} \tag{10-6}$$

and

$$\begin{cases} x = x_iL_i + x_jL_j + x_kL_k \\ y = y_iL_i + y_jL_j + y_kL_k \end{cases} \tag{10-7}$$

The displacements for triangle element

$$\begin{cases} u = N_iu_i + N_ju_j + N_ku_k \\ v = N_iv_i + N_jv_j + N_kv_k \end{cases} \tag{10-8}$$

It can be seen that for triangle element, we also use the same shape function to express the coordinates and the displacements.

10.3 Quadrilateral element

For a quadrilateral element, according to Pascal triangle rule, the displacement can be expressed as

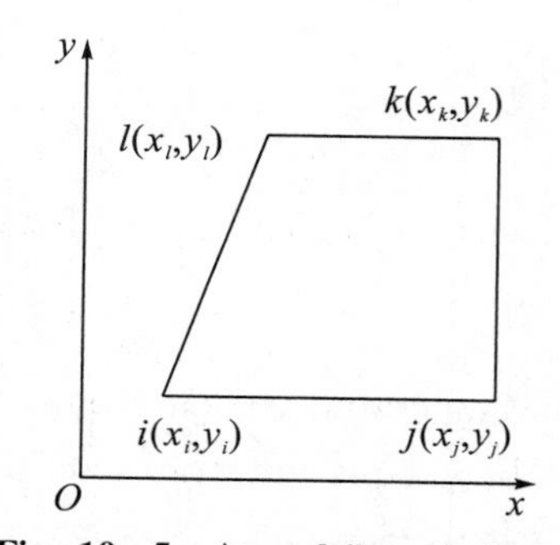

Fig. 10-5 A quadrilateral element

$$\begin{cases} u = \alpha_1 + \alpha_2x + \alpha_3y + \alpha_4xy \\ v = \alpha_5 + \alpha_6x + \alpha_7y + \alpha_8xy \end{cases} \tag{10-9}$$

The displacements on the mutual boundary may be not compatible, but after mapped onto the standard square element, which doesn't have such problem. The displacements can be rewritten as

$$\begin{cases} u = N_iu_i + N_ju_j + N_ku_k + N_lu_l \\ v = N_iv_i + N_jv_j + N_kv_k + N_lv_l \end{cases} \tag{10-10}$$

where N_i, N_j, N_k and N_l are the shape functions of the square element, which is presented in Eq. (10-4).

According to the relationship between strain and displacement, one can have

$$\{\varepsilon\}=\begin{Bmatrix}\varepsilon_x\\ \varepsilon_y\\ \gamma_{xy}\end{Bmatrix}=\begin{Bmatrix}\dfrac{\partial u}{\partial x}\\ \dfrac{\partial v}{\partial y}\\ \dfrac{\partial u}{\partial y}+\dfrac{\partial v}{\partial x}\end{Bmatrix}=\begin{bmatrix}\dfrac{\partial}{\partial x} & 0\\ 0 & \dfrac{\partial}{\partial y}\\ \dfrac{\partial}{\partial y} & \dfrac{\partial}{\partial x}\end{bmatrix}\begin{Bmatrix}u\\ v\end{Bmatrix} \tag{10-11}$$

Substituting the displacements into Eq. (10—11), one can have

$$\{\varepsilon\}=\begin{bmatrix}\dfrac{\partial N_i}{\partial x} & 0 & \dfrac{\partial N_j}{\partial x} & 0 & \dfrac{\partial N_k}{\partial x} & 0 & \dfrac{\partial N_l}{\partial x} & 0\\ 0 & \dfrac{\partial N_i}{\partial y} & 0 & \dfrac{\partial N_j}{\partial y} & 0 & \dfrac{\partial N_k}{\partial y} & 0 & \dfrac{\partial N_l}{\partial y}\\ \dfrac{\partial N_i}{\partial y} & \dfrac{\partial N_i}{\partial x} & \dfrac{\partial N_j}{\partial y} & \dfrac{\partial N_j}{\partial x} & \dfrac{\partial N_k}{\partial y} & \dfrac{\partial N_k}{\partial x} & \dfrac{\partial N_l}{\partial y} & \dfrac{\partial N_l}{\partial x}\end{bmatrix}\begin{Bmatrix}u_i\\ v_i\\ u_j\\ v_j\\ u_k\\ v_k\\ u_l\\ v_l\end{Bmatrix}=[B]\{\delta\}^e \tag{10-12}$$

According to the relationship between stress and strain, the stress can be expressed as

$$[\sigma]=[D][B]\{\delta\}^e \tag{10-13}$$

where $[D]$ is plane stress elasticity matrix.

$$[D]=\frac{E}{1-\nu^2}\begin{bmatrix}1 & \nu & 0\\ \nu & 1 & 0\\ 0 & 0 & \dfrac{1-\nu}{2}\end{bmatrix} \tag{10-14}$$

According to the principle of virtual displacement, for an element, one can have

$$(\{\delta^*\}^e)^{\mathrm{T}}\{F\}^e=\iint_A\{\varepsilon^*\}^{\mathrm{T}}\{\sigma\}t\,\mathrm{d}x\,\mathrm{d}y \tag{10-15}$$

where δ^* and ε^* are virtual displacement and virtual strain, respectively. Substituting the stress and strain matrix into the above equation, one can have

$$\{F\}^e=\left(\iint_A[B]^{\mathrm{T}}[D][B]t\,\mathrm{d}x\,\mathrm{d}y\right)\{\delta\}^e \tag{10-16}$$

The element stiffness matrix is

$$[K]^e=\iint[B]^{\mathrm{T}}[D][B]\mathrm{d}x\,\mathrm{d}y \tag{10-17}$$

where $[B]$ is given by

$$[B]=\begin{bmatrix}\dfrac{\partial N_i}{\partial x} & 0 & \dfrac{\partial N_j}{\partial x} & 0 & \dfrac{\partial N_k}{\partial x} & 0 & \dfrac{\partial N_l}{\partial x} & 0\\ 0 & \dfrac{\partial N_i}{\partial y} & 0 & \dfrac{\partial N_j}{\partial y} & 0 & \dfrac{\partial N_k}{\partial y} & 0 & \dfrac{\partial N_l}{\partial y}\\ \dfrac{\partial N_i}{\partial y} & \dfrac{\partial N_i}{\partial x} & \dfrac{\partial N_j}{\partial y} & \dfrac{\partial N_j}{\partial x} & \dfrac{\partial N_k}{\partial y} & \dfrac{\partial N_k}{\partial x} & \dfrac{\partial N_l}{\partial y} & \dfrac{\partial N_l}{\partial x}\end{bmatrix} \tag{10-18}$$

It can be seen that the $[B]$ matrix contains the partial derivative of the shape function with respect to x, y coordinates, and the integration in Eq. (10-17) is with respect to the area $dxdy$. Since we have to use local coordinates (ξ, η) to calculate the element stiffness, and therefore, it is necessary to devise some means of expressing the global derivatives in terms of local derivatives; i. e. using $\frac{\partial N_i}{\partial \xi}$, $\frac{\partial N_i}{\partial \eta}$ to replace $\frac{\partial N_i}{\partial x}$, $\frac{\partial N_i}{\partial y}$, and using $d\xi d\eta$ to replace $dxdy$ in the integration in Eq. (10-17).

10.4 Relationship between $\frac{\partial N_i}{\partial \xi}$, $\frac{\partial N_i}{\partial \eta}$, $\frac{\partial N_i}{\partial \zeta}$ and $\frac{\partial N_i}{\partial x}$, $\frac{\partial N_i}{\partial y}$, $\frac{\partial N_i}{\partial z}$

Consider the set of local coordinates ξ, η, ζ and a corresponding set of global coordinates x, y, z. By the usual rules of partial differentiation we can write, for instance, the ξ derivative as

$$\frac{\partial N_i}{\partial \xi}=\frac{\partial N_i}{\partial x}\cdot\frac{\partial x}{\partial \xi}+\frac{\partial N_i}{\partial y}\cdot\frac{\partial y}{\partial \xi}+\frac{\partial N_i}{\partial z}\cdot\frac{\partial z}{\partial \xi} \tag{10-19}$$

Performing the same differentiation with respect to the other two coordinates and writing in matrix form, we have

$$\begin{Bmatrix}\frac{\partial N_i}{\partial \xi}\\ \frac{\partial N_i}{\partial \eta}\\ \frac{\partial N_i}{\partial \zeta}\end{Bmatrix}=\begin{bmatrix}\frac{\partial x}{\partial \xi} & \frac{\partial y}{\partial \xi} & \frac{\partial z}{\partial \xi}\\ \frac{\partial x}{\partial \eta} & \frac{\partial y}{\partial \eta} & \frac{\partial z}{\partial \eta}\\ \frac{\partial x}{\partial \zeta} & \frac{\partial y}{\partial \zeta} & \frac{\partial z}{\partial \zeta}\end{bmatrix}\begin{Bmatrix}\frac{\partial N_i}{\partial x}\\ \frac{\partial N_i}{\partial y}\\ \frac{\partial N_i}{\partial z}\end{Bmatrix}=\boldsymbol{J}\begin{Bmatrix}\frac{\partial N_i}{\partial x}\\ \frac{\partial N_i}{\partial y}\\ \frac{\partial N_i}{\partial z}\end{Bmatrix} \tag{10-20}$$

where $\boldsymbol{J}$ is the well known Jacobian matrix.

To find now the derivativeswith respect to x, y, we invert $\boldsymbol{J}$ and we have

$$\begin{Bmatrix}\frac{\partial N_i}{\partial x}\\ \frac{\partial N_i}{\partial y}\\ \frac{\partial N_i}{\partial z}\end{Bmatrix}=\boldsymbol{J}^{-1}\begin{Bmatrix}\frac{\partial N_i}{\partial \xi}\\ \frac{\partial N_i}{\partial \eta}\\ \frac{\partial N_i}{\partial \zeta}\end{Bmatrix} \tag{10-21}$$

where $\boldsymbol{J}^{-1}=\frac{1}{|\boldsymbol{J}|}\boldsymbol{J}^*$.

According to the knowledge of Linear Algebra, if $\boldsymbol{J}$ is written as

$$\boldsymbol{J}=\begin{bmatrix}a_{11} & a_{12} & a_{13}\\ a_{21} & a_{22} & a_{23}\\ a_{31} & a_{32} & a_{33}\end{bmatrix} \tag{10-22}$$

then

$$J^* = \begin{bmatrix} A_{11} & A_{21} & A_{31} \\ A_{12} & A_{22} & A_{32} \\ A_{13} & A_{23} & A_{33} \end{bmatrix} \tag{10-23}$$

where A_{ij} can be calculated. For example, A_{12} can be calculated by

$$A_{12} = (-1)^{i+j} \begin{vmatrix} a_{21} & a_{23} \\ a_{31} & a_{33} \end{vmatrix} = -(a_{21}a_{33} - a_{31}a_{23}) \tag{10-24}$$

For example

$$J = \begin{bmatrix} 2 & 2 & 2 \\ 1 & 2 & 3 \\ 1 & 3 & 6 \end{bmatrix}$$

The determinant $|J| = 2$, and

$$J^* = \begin{bmatrix} 3 & -6 & 2 \\ -3 & 10 & -4 \\ 1 & -4 & 2 \end{bmatrix}$$

Finally

$$J^{-1} = \frac{1}{2}\begin{bmatrix} 3 & -6 & 2 \\ -3 & 10 & -4 \\ 1 & -4 & 2 \end{bmatrix}$$

For 2D case, the Jacobian matrix can be simplified as

$$J = \begin{bmatrix} \frac{\partial x}{\partial \xi} & \frac{\partial y}{\partial \xi} \\ \frac{\partial x}{\partial \eta} & \frac{\partial y}{\partial \eta} \end{bmatrix} \tag{10-25}$$

The calculation method of the Jacobian matrix is introduced in the follows. The Cartesian coordinates x, y can be expressed as

$$\begin{cases} x = N_1 x_1 + N_2 x_2 + N_3 x_3 + N_4 x_4 \\ y = N_1 y_1 + N_2 y_2 + N_3 y_3 + N_4 y_4 \end{cases} \tag{10-26}$$

Substituting Eq. (10−26) into Eq. (10−25), for 2D case, we have

$$J = \begin{bmatrix} \frac{\partial N_1}{\partial \xi}x_1 + \frac{\partial N_2}{\partial \xi}x_2 + \frac{\partial N_3}{\partial \xi}x_3 + \frac{\partial N_4}{\partial \xi}x_4 & \frac{\partial N_1}{\partial \xi}y_1 + \frac{\partial N_2}{\partial \xi}y_2 + \frac{\partial N_3}{\partial \xi}y_3 + \frac{\partial N_4}{\partial \xi}y_4 \\ \frac{\partial N_1}{\partial \eta}x_1 + \frac{\partial N_2}{\partial \eta}x_2 + \frac{\partial N_3}{\partial \eta}x_3 + \frac{\partial N_4}{\partial \eta}x_4 & \frac{\partial N_1}{\partial \eta}y_1 + \frac{\partial N_2}{\partial \eta}y_2 + \frac{\partial N_3}{\partial \eta}y_3 + \frac{\partial N_4}{\partial \eta}y_4 \end{bmatrix}$$

$$= \begin{bmatrix} \frac{\partial N_1}{\partial \xi} & \frac{\partial N_2}{\partial \xi} & \frac{\partial N_3}{\partial \xi} & \frac{\partial N_4}{\partial \xi} \\ \frac{\partial N_1}{\partial \eta} & \frac{\partial N_2}{\partial \eta} & \frac{\partial N_3}{\partial \eta} & \frac{\partial N_4}{\partial \eta} \end{bmatrix} \begin{bmatrix} x_1 & y_1 \\ x_2 & y_2 \\ x_3 & y_3 \\ x_4 & y_4 \end{bmatrix} \tag{10-27}$$

For example, a four-node distorted element as shown in Fig. 10−6, where the length of $O1$, $O2$, $O3$ and $O4$ is 1.0, calculate the Jacobian matrix, and calculate the derivative of shape function with respect to the x, y coordinates $\begin{Bmatrix} \dfrac{\partial N_k}{\partial x} \\ \dfrac{\partial N_k}{\partial y} \end{Bmatrix}$.

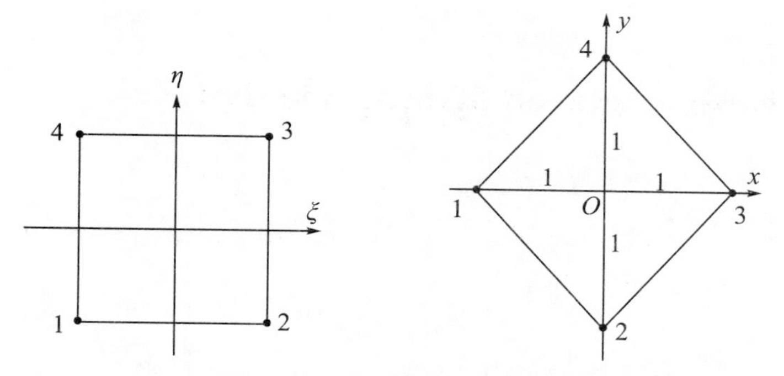

Fig. 10−6 A standard square element and a distorted element

The shape functions can be written as

$$N_1 = \frac{1}{4}(1-\xi)(1-\eta)$$

$$N_2 = \frac{1}{4}(1+\xi)(1-\eta)$$

$$N_3 = \frac{1}{4}(1+\xi)(1+\eta)$$

$$N_4 = \frac{1}{4}(1-\xi)(1+\eta)$$

Substituting them into Eq. (10−27), we have

$$\boldsymbol{J} = \begin{bmatrix} \dfrac{\partial x}{\partial \xi} & \dfrac{\partial y}{\partial \xi} \\ \dfrac{\partial x}{\partial \eta} & \dfrac{\partial y}{\partial \eta} \end{bmatrix} = \begin{bmatrix} \dfrac{\partial N_1}{\partial \xi} & \dfrac{\partial N_2}{\partial \xi} & \dfrac{\partial N_3}{\partial \xi} & \dfrac{\partial N_4}{\partial \xi} \\ \dfrac{\partial N_1}{\partial \eta} & \dfrac{\partial N_2}{\partial \eta} & \dfrac{\partial N_3}{\partial \eta} & \dfrac{\partial N_4}{\partial \eta} \end{bmatrix} \begin{bmatrix} x_1 & y_1 \\ x_2 & y_2 \\ x_3 & y_3 \\ x_4 & y_4 \end{bmatrix}$$

$$= \frac{1}{4}\begin{bmatrix} -(1-\eta) & 1-\eta & 1+\eta & -(1+\eta) \\ -(1-\xi) & -(1+\xi) & 1+\xi & 1-\xi \end{bmatrix} \begin{bmatrix} -1 & 0 \\ 0 & -1 \\ 1 & 0 \\ 0 & 1 \end{bmatrix} = \begin{bmatrix} 0.5 & -0.5 \\ 0.5 & 0.5 \end{bmatrix}$$

The partial derivative of the shape function with respect to x, y can be expressed as

$$\begin{Bmatrix} \dfrac{\partial N_i}{\partial x} \\ \dfrac{\partial N_i}{\partial y} \end{Bmatrix} = \boldsymbol{J}^{-1} \begin{Bmatrix} \dfrac{\partial N_i}{\partial \xi} \\ \dfrac{\partial N_i}{\partial \eta} \end{Bmatrix}$$

where

$$\boldsymbol{J}^{-1} = \frac{1}{|A|}A^* = \frac{1}{2}\begin{bmatrix} 0.5 & -0.5 \\ 0.5 & 0.5 \end{bmatrix} = \begin{bmatrix} 1 & 1 \\ -1 & 1 \end{bmatrix}$$

$$\begin{Bmatrix} \dfrac{\partial N_k}{\partial x} \\ \dfrac{\partial N_k}{\partial y} \end{Bmatrix} = \boldsymbol{J}^{-1} \begin{Bmatrix} \dfrac{\partial N_k}{\partial \xi} \\ \dfrac{\partial N_k}{\partial \eta} \end{Bmatrix} = \begin{bmatrix} 1 & 1 \\ -1 & 1 \end{bmatrix} \begin{Bmatrix} \dfrac{\partial N_k}{\partial \xi} \\ \dfrac{\partial N_k}{\partial \eta} \end{Bmatrix} = \begin{bmatrix} 1 & 1 \\ -1 & 1 \end{bmatrix} \frac{1}{4} \begin{Bmatrix} 1+\eta \\ 1+\xi \end{Bmatrix} = \frac{1}{4} \begin{Bmatrix} 2+\xi+\eta \\ \xi-\eta \end{Bmatrix}$$

10.5 Relationship between $\mathrm{d}\xi\mathrm{d}\eta\mathrm{d}\zeta$ and $\mathrm{d}x\mathrm{d}y\mathrm{d}z$

For 3D element, as shown in Fig. 10−7, the volume of the vectors, $\mathrm{d}x$, $\mathrm{d}y$, $\mathrm{d}z$, in the global coordinates can be written as

$$\mathrm{d}V = \mathrm{d}x \cdot \mathrm{d}y \times \mathrm{d}z \tag{10-28}$$

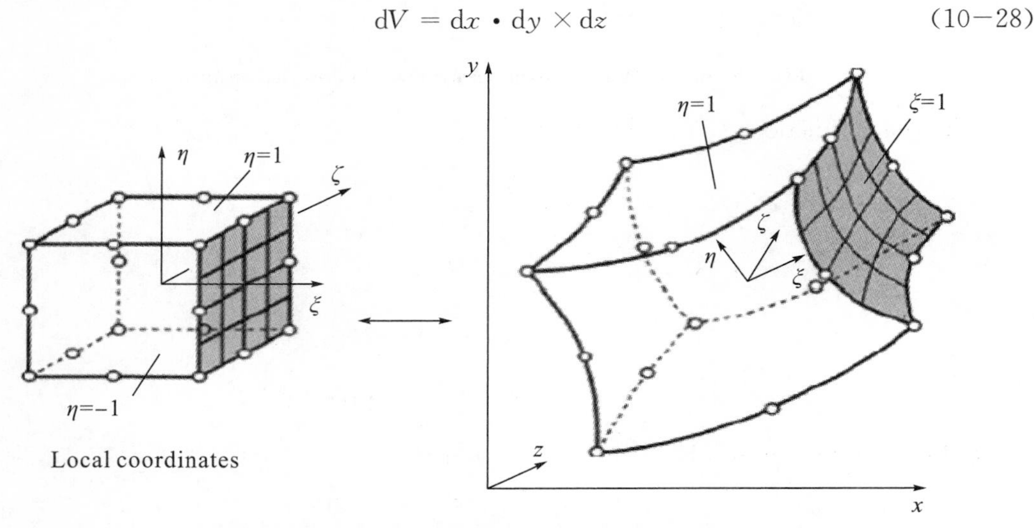

Fig. 10−7 A standard cubic element and a distorted element

The relationship between $\mathrm{d}x$, $\mathrm{d}y$, $\mathrm{d}z$ and $\mathrm{d}\xi$, $\mathrm{d}\eta$, $\mathrm{d}\zeta$ can be expressed as

$$\begin{cases} \mathrm{d}x = \dfrac{\partial x}{\partial \xi}\mathrm{d}\xi i + \dfrac{\partial x}{\partial \eta}\mathrm{d}\eta j + \dfrac{\partial x}{\partial \zeta}\mathrm{d}\zeta k \\ \mathrm{d}y = \dfrac{\partial y}{\partial \xi}\mathrm{d}\xi i + \dfrac{\partial y}{\partial \eta}\mathrm{d}\eta j + \dfrac{\partial y}{\partial \zeta}\mathrm{d}\zeta k \\ \mathrm{d}z = \dfrac{\partial z}{\partial \xi}\mathrm{d}\xi i + \dfrac{\partial z}{\partial \eta}\mathrm{d}\eta j + \dfrac{\partial z}{\partial \zeta}\mathrm{d}\zeta k \end{cases} \tag{10-29}$$

and

$$\mathrm{d}x\mathrm{d}y\mathrm{d}z = \begin{vmatrix} \dfrac{\partial x}{\partial \xi} & \dfrac{\partial x}{\partial \eta} & \dfrac{\partial x}{\partial \zeta} \\ \dfrac{\partial y}{\partial \xi} & \dfrac{\partial y}{\partial \eta} & \dfrac{\partial y}{\partial \zeta} \\ \dfrac{\partial z}{\partial \xi} & \dfrac{\partial z}{\partial \eta} & \dfrac{\partial z}{\partial \zeta} \end{vmatrix} \mathrm{d}\xi\mathrm{d}\eta\mathrm{d}\zeta = |\boldsymbol{J}|\,\mathrm{d}\xi\mathrm{d}\eta\mathrm{d}\zeta \tag{10-30}$$

where

$$J=\begin{Bmatrix}\frac{\partial x}{\partial \xi} & \frac{\partial y}{\partial \xi} & \frac{\partial z}{\partial \xi}\\ \frac{\partial x}{\partial \eta} & \frac{\partial y}{\partial \eta} & \frac{\partial z}{\partial \eta}\\ \frac{\partial x}{\partial \zeta} & \frac{\partial y}{\partial \zeta} & \frac{\partial z}{\partial \zeta}\end{Bmatrix}$$

It should be noted that $|\boldsymbol{J}^{\mathrm{T}}|=|\boldsymbol{J}|$.

For 2D case, Eq. (10−29) can be simplified as

$$\begin{cases}\mathrm{d}x=\frac{\partial x}{\partial \xi}\mathrm{d}\xi i+\frac{\partial x}{\partial \eta}\mathrm{d}\eta j\\ \mathrm{d}y=\frac{\partial y}{\partial \xi}\mathrm{d}\xi i+\frac{\partial y}{\partial \eta}\mathrm{d}\eta j\end{cases} \tag{10−31}$$

The corresponding volume integral changes to area integral as

$$\mathrm{d}x\mathrm{d}y=|\boldsymbol{J}|\mathrm{d}\xi\mathrm{d}\eta \tag{10−32}$$

Substituting Eq. (10−32) into Eq. (10−17), the element stiffness can be expressed as

$$[K]^{e}=\iint[B]^{\mathrm{T}}[D][B]\mathrm{d}x\mathrm{d}y=\iint[B]^{\mathrm{T}}[D][B]|\boldsymbol{J}|\mathrm{d}\xi\mathrm{d}\eta \tag{10−33}$$

For 3D case, the element stiffness can be expressed as

$$[K]^{e}=\iiint[B]^{\mathrm{T}}[D][B]\mathrm{d}x\mathrm{d}y\mathrm{d}z=\iiint[B]^{\mathrm{T}}[D][B]|\boldsymbol{J}|\mathrm{d}\xi\mathrm{d}\eta\mathrm{d}\zeta \tag{10−34}$$

Finally, using $\frac{\partial N_i}{\partial \xi}$, $\frac{\partial N_i}{\partial \eta}$ to replace $\frac{\partial N_i}{\partial x}$, $\frac{\partial N_i}{\partial y}$, and using $\mathrm{d}\xi\mathrm{d}\eta$ to replace $\mathrm{d}x\mathrm{d}y$, the integration of the element stiffness is only related to the ξ, η (and ζ for 3D) coordinates, and the corresponding element is a standard square element. The displacements on the mutual boundary are, therefore, compatible.

10.6 Discussion

According to the calculus theory, the condition of one-to-one transformation between two coordinates is that the determinant of the Jacobian matrix is not zero, i. e. $|J|\neq 0$, that means the transformation between the global coordinates and the local coordinates must obey the condition $|\boldsymbol{J}|\neq 0$.

If $|\boldsymbol{J}|\neq 0$, $\mathrm{d}x\mathrm{d}y=|\boldsymbol{J}|\mathrm{d}\xi\mathrm{d}\eta=0$, that means a small area in the local coordinates corresponds to only one point in the global coordinates, and it is obviously not a one-to-one transformation. Meanwhile, as $|\boldsymbol{J}|\neq 0$, the inverse of Jacobian matrix does exist.

For the four-node quadrilateral element as shown in Fig. 10−8, the triangular part 123 in the standard element cannot map into the part 123 of the distorted elements for the case 1 and 2. This is because the area of triangle 123 will be mapped into zero area in case

1 and negative area in case 2, that are obviously not one-to-one transformation. Therefore, those four-node element cannot be used in the isoparametric element method.

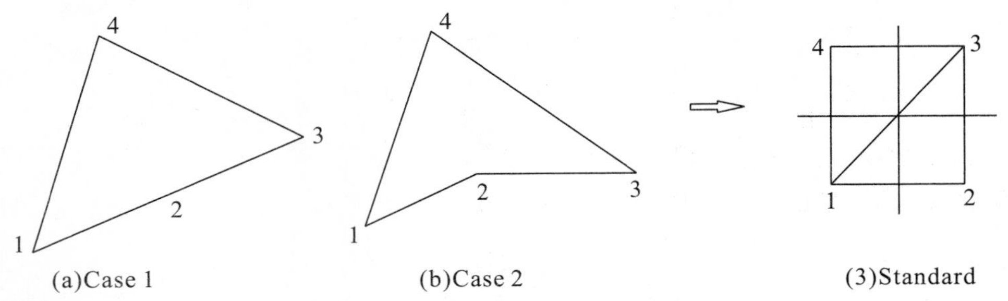

Fig. 10−8 Two four-node elements

10.7 Some distorted elements

The 2D quadrilateral element with eight-node, the cubic element with eight-node and with twenty-node will be discussed because they are used widely.

10.7.1 Quadrilateral element with eight-node

For the eight-node distorted element as shown in Fig. 10−9 (a), one has to use the mapping technique to study it. We take the displacement function for the standard element as shown in Fig. 10−9 (b) as

$$\begin{cases} u = \alpha_1 + \alpha_2\xi + \alpha_3\eta + \alpha_4\xi^2 + \alpha_5\xi\eta + \alpha_6\eta^2 + \alpha_7\xi^2\eta + \alpha_8\xi\eta^2 \\ v = \alpha_9 + \alpha_{10}\xi + \alpha_{11}\eta + \alpha_{12}\xi^2 + \alpha_{13}\xi\eta + \alpha_{14}\eta^2 + \alpha_{15}\xi^2\eta + \alpha_{16}\xi\eta^2 \end{cases} \tag{10-35}$$

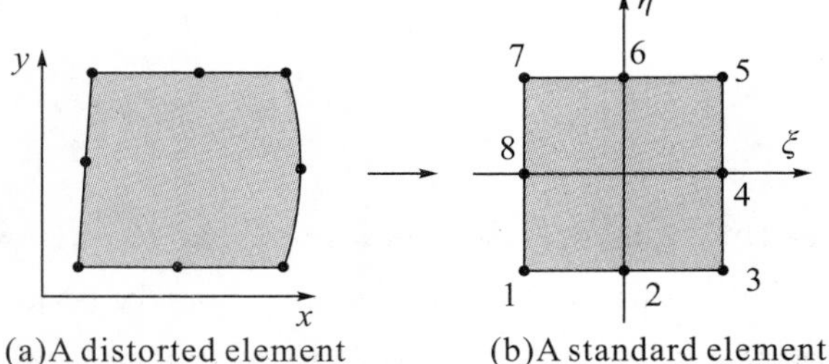

Fig. 10−9 A distorted element and a standard element

Discussion: for the standard eight-node element shown in Fig. 10−9 (b), the displacement on each edge can be written as $u = C_1 + C_2\eta + C_3\eta^2$ or $v = C_1 + C_2\xi + C_3\xi^2$. Because there are three nodes on each edge and three coefficients in the equation, the curve of the displacement on each edge is fixed. On the mutual edge ABC of the two adjacent element shown in Fig. 10−10, because for both element (1) and (2), the displacements of nodes A, B and C are the same and fixed, the curves in both sides are exactly the same. Therefore, the displacements occurring in the mutual boundary ABC of the two adjacent elements are equal, which can make the boundary continuous and compatible, avoiding dislocations.

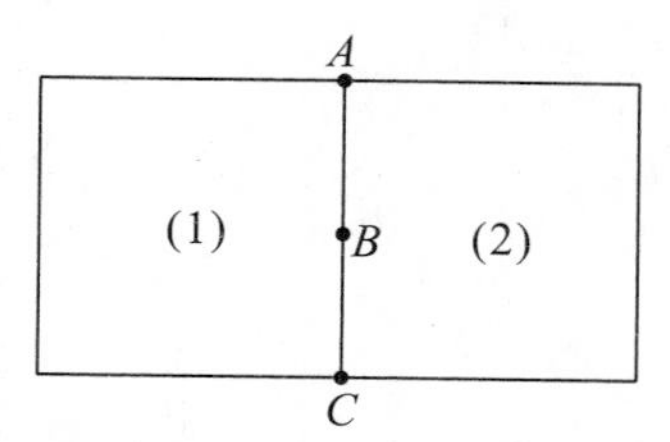

Fig. 10—10 Two elements

From Eq. (10—35), the corresponding shape function, which will be used to express the x, y coordinates and the displacements of the distorted element, can be obtained as

$$N_i(\xi,\eta)=\begin{cases}\dfrac{1}{4}(1+\xi_i\xi)(1+\eta_i\eta)(\xi_i\xi+\eta_i\eta-1) & (i=1,3,5,7)\\ \dfrac{1}{2}(1-\xi^2)(1+\eta_i\eta) & (i=2,6)\\ \dfrac{1}{2}(1-\eta^2)(1+\xi_i\xi) & (i=4,8)\end{cases} \tag{10—36}$$

The displacements of the distorted element can be expressed in terms of shape function as

$$\begin{cases}u=\sum\limits_{i=1}^{8}N_i(\xi,\eta)u_i\\ v=\sum\limits_{i=1}^{8}N_i(\xi,\eta)v_i\end{cases} \tag{10—37}$$

The coordinate transformation can be written as

$$\begin{cases}x=\sum\limits_{i=1}^{8}N_i(\xi,\eta)x_i\\ y=\sum\limits_{i=1}^{8}N_i(\xi,\eta)y_i\end{cases} \tag{10—38}$$

The relationship between strains and stresses can be expressed as

$$\{\varepsilon\}=\begin{Bmatrix}\varepsilon_x\\ \varepsilon_y\\ \gamma_{xy}\end{Bmatrix}=\begin{Bmatrix}\dfrac{\partial u}{\partial x}\\ \dfrac{\partial v}{\partial y}\\ \dfrac{\partial u}{\partial y}+\dfrac{\partial v}{\partial x}\end{Bmatrix}=\begin{bmatrix}\dfrac{\partial}{\partial x} & 0\\ 0 & \dfrac{\partial}{\partial y}\\ \dfrac{\partial}{\partial y} & \dfrac{\partial}{\partial x}\end{bmatrix}\begin{Bmatrix}u\\ v\end{Bmatrix} \tag{10—39}$$

Substituting the displacements into Eq. (10—39), one can have

$$[\varepsilon]=\begin{Bmatrix}\varepsilon_x\\ \varepsilon_y\\ \gamma_{xy}\end{Bmatrix}=\begin{bmatrix}\frac{\partial N_1}{\partial x} & 0 & \frac{\partial N_2}{\partial x} & 0 & \frac{\partial N_3}{\partial x} & 0 & \cdots & \frac{\partial N_8}{\partial x} & 0\\ 0 & \frac{\partial N_1}{\partial y} & 0 & \frac{\partial N_2}{\partial y} & 0 & \frac{\partial N_3}{\partial y} & \cdots & 0 & \frac{\partial N_8}{\partial y}\\ \frac{\partial N_1}{\partial y} & \frac{\partial N_1}{\partial x} & \frac{\partial N_2}{\partial y} & \frac{\partial N_2}{\partial x} & \frac{\partial N_3}{\partial y} & \frac{\partial N_3}{\partial x} & \cdots & \frac{\partial N_8}{\partial y} & \frac{\partial N_8}{\partial x}\end{bmatrix}\begin{Bmatrix}\delta_{1x}\\ \delta_{1y}\\ \delta_{2x}\\ \delta_{2y}\\ \vdots\\ \delta_{8x}\\ \delta_{8y}\end{Bmatrix} \tag{10-40}$$

The Jacobian matrix can be calculated by

$$\boldsymbol{J}=\begin{bmatrix}\frac{\partial x}{\partial \xi} & \frac{\partial y}{\partial \xi}\\ \frac{\partial x}{\partial \eta} & \frac{\partial y}{\partial \eta}\end{bmatrix}=\begin{bmatrix}\sum \frac{\partial N_i}{\partial \xi}x_i & \sum \frac{\partial N_i}{\partial \xi}y_i\\ \sum \frac{\partial N_i}{\partial \eta}x_i & \sum \frac{\partial N_i}{\partial \eta}y_i\end{bmatrix}=\begin{bmatrix}\frac{\partial N_1}{\partial \xi} & \frac{\partial N_2}{\partial \xi} & \cdots & \frac{\partial N_8}{\partial \xi}\\ \frac{\partial N_1}{\partial \eta} & \frac{\partial N_2}{\partial \eta} & \cdots & \frac{\partial N_8}{\partial \eta}\end{bmatrix}\begin{bmatrix}x_1 & y_1\\ x_2 & y_2\\ \vdots & \vdots\\ x_8 & y_8\end{bmatrix} \tag{10-41}$$

Using $\frac{\partial N_i}{\partial \xi}$, $\frac{\partial N_i}{\partial \eta}$ to replace $\frac{\partial N_i}{\partial x}$, $\frac{\partial N_i}{\partial y}$, i. e.

$$\begin{Bmatrix}\frac{\partial N_i}{\partial x}\\ \frac{\partial N_i}{\partial y}\end{Bmatrix}=\boldsymbol{J}^{-1}\begin{Bmatrix}\frac{\partial N_i}{\partial \xi}\\ \frac{\partial N_i}{\partial \eta}\end{Bmatrix}$$

Finally substituting the shape function in Eq. (10−36) into Eq. (10−40), the $[B]$ matrix can be obtained which contains ξ, η. Eq. (10−40) can be simplified as

$$\{\varepsilon\}=[B]\{\delta\}^e \tag{10-42}$$

For plane problems, the element stiffness can be written as

$$[K]^e=\int_{-1}^{1}\int_{-1}^{1}[B]^{\mathrm{T}}[D][B]t\,|\boldsymbol{J}|\,\mathrm{d}\xi\mathrm{d}\eta \tag{10-43}$$

The element stiffness matrix contains the coordinates ξ, η, and the $[K]^e$ can be calculated by using the Gauss integration method, which will be studied in next chapter.

$$[K]^e=\begin{bmatrix}K_{11} & K_{12} & K_{13} & \cdots & K_{18}\\ K_{21} & K_{22} & K_{23} & \cdots & K_{28}\\ K_{31} & K_{32} & K_{33} & \cdots & K_{38}\\ \vdots & \vdots & \vdots & & \vdots\\ K_{81} & K_{82} & K_{83} & \cdots & K_{88}\end{bmatrix} \tag{10-44}$$

In Eq. (10−44), the element stiffness contains 256 (16×16) coefficients because each term in Eq. (10−44) has four coefficients.

For the structure shown in Fig. 10−11, the stiffness of element (1) should be installed into the global stiffness matrix as follows

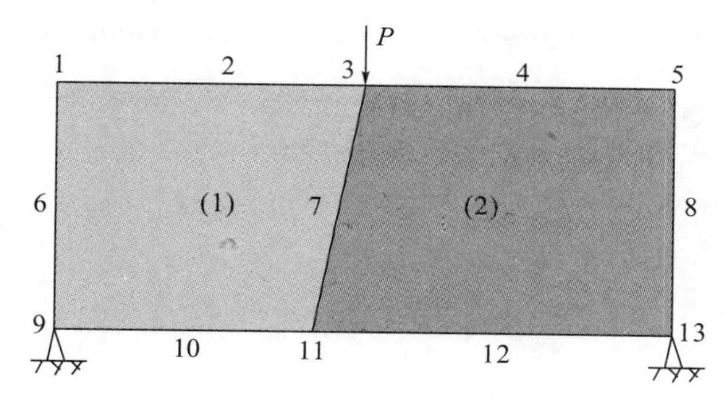

Fig. 10−11 Two distorted elements

$$
\begin{bmatrix}
K_{11} & K_{12} & K_{13} & & & K_{16} & K_{17} & & K_{19} & K_{110} & K_{111} & & \\
K_{21} & K_{22} & K_{23} & & & K_{26} & K_{27} & & K_{29} & K_{210} & K_{211} & & \\
K_{31} & K_{32} & K_{33} & & & K_{36} & K_{37} & & K_{39} & K_{310} & K_{311} & & \\
 & & & & & & & & & & & & \\
 & & & & & & & & & & & & \\
K_{61} & K_{62} & K_{63} & & & K_{66} & K_{67} & & K_{69} & K_{610} & K_{611} & & \\
K_{71} & K_{72} & K_{73} & & & K_{76} & K_{77} & & K_{79} & K_{710} & K_{711} & & \\
 & & & & & & & & & & & & \\
K_{91} & K_{92} & K_{93} & & & K_{96} & K_{97} & & K_{99} & K_{910} & K_{911} & & \\
K_{101} & K_{102} & K_{103} & & & K_{106} & K_{107} & & K_{109} & K_{1010} & K_{1011} & & \\
K_{111} & K_{112} & K_{113} & & & K_{116} & K_{117} & & K_{119} & K_{1110} & K_{1111} & & \\
 & & & & & & & & & & & & \\
 & & & & & & & & & & & &
\end{bmatrix}
\begin{Bmatrix} \delta_1 \\ \delta_2 \\ \delta_3 \\ \delta_4 \\ \delta_5 \\ \delta_6 \\ \delta_7 \\ \delta_8 \\ \delta_9 \\ \delta_{10} \\ \delta_{11} \\ \delta_{12} \\ \delta_{13} \end{Bmatrix}
\begin{Bmatrix} F_1 \\ F_2 \\ F_3 \\ F_4 \\ F_5 \\ F_6 \\ F_7 \\ F_8 \\ F_9 \\ F_{10} \\ F_{11} \\ F_{12} \\ F_{13} \end{Bmatrix}
\tag{10−45}
$$

10.7.2 Eight-node 3D isoparametric element

The eight-node distorted element and the corresponding standard element are shown in Fig. 10−12. The displacement function of the standard element can be written as

$$
\begin{cases}
u = \alpha_1 + \alpha_2\xi + \alpha_3\eta + \alpha_4\zeta + \alpha_5\xi\eta + \alpha_6\eta\xi + \alpha_7\xi\zeta + \alpha_8\xi\eta\zeta \\
v = \alpha_9 + \alpha_{10}\xi + \alpha_{11}\eta + \alpha_{12}\zeta + \alpha_{13}\xi\eta + \alpha_{14}\eta\xi + \alpha_{15}\xi\zeta + \alpha_{16}\xi\eta\zeta \\
w = \alpha_{17} + \alpha_{18}\xi + \alpha_{19}\eta + \alpha_{20}\zeta + \alpha_{21}\xi\eta + \alpha_{22}\eta\xi + \alpha_{23}\xi\zeta + \alpha_{24}\xi\eta\zeta
\end{cases}
\tag{10−46}
$$

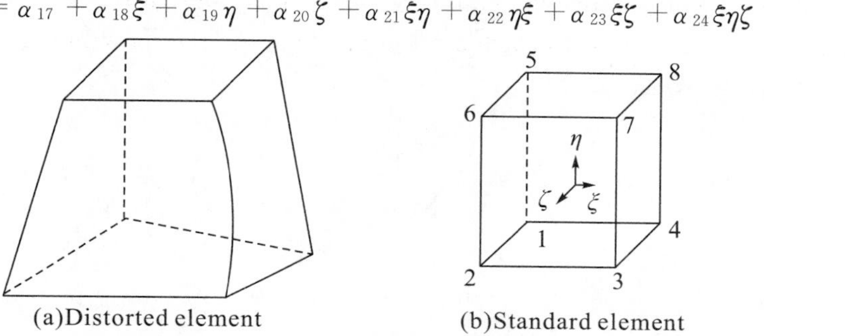

Fig. 10−12 A distorted element and the standard element

The displacement can be expressed in terms of shape function as

$$\begin{cases} u = \sum_{i=1}^{8} N_i(\xi, \eta, \zeta)u_i \\ v = \sum_{i=1}^{8} N_i(\xi, \eta, \zeta)v_i \\ w = \sum_{i=1}^{8} N_i(\xi, \eta, \zeta)w_i \end{cases} \tag{10-47}$$

where

$$N_i(\xi,\eta,\zeta) = \frac{1}{8}(1+\xi_i\xi)(1+\eta_i\eta)(1+\zeta_i\zeta) \quad (i=1, 2, \cdots, 8) \tag{10-48}$$

The coordinate transformation can be written as

$$\begin{cases} x = \sum_{i=1}^{8} N_i(\xi, \eta, \zeta)x_i \\ y = \sum_{i=1}^{8} N_i(\xi, \eta, \zeta)y_i \\ z = \sum_{i=1}^{8} N_i(\xi, \eta, \zeta)z_i \end{cases} \tag{10-49}$$

The relationship of trains versus displacements can be expressed as

$$\{\varepsilon\} = \begin{Bmatrix} \varepsilon_x \\ \varepsilon_y \\ \varepsilon_z \\ \gamma_{xy} \\ \gamma_{yz} \\ \gamma_{zx} \end{Bmatrix} = \begin{Bmatrix} \frac{\partial u}{\partial x} \\ \frac{\partial v}{\partial y} \\ \frac{\partial w}{\partial z} \\ \frac{\partial u}{\partial y} + \frac{\partial v}{\partial x} \\ \frac{\partial v}{\partial z} + \frac{\partial w}{\partial y} \\ \frac{\partial w}{\partial x} + \frac{\partial u}{\partial z} \end{Bmatrix} \tag{10-50}$$

Substituting the displacements in Eq. (10−47) into Eq. (10−50), one can obtain the strain matrix [B] which contains the partial derivative with respect to x, y coordinates. We first need to calculate the Jacobian matrix shown in Eq. (10−31) and its reciprocal, then using

$$\begin{Bmatrix} \frac{\partial N_i}{\partial x} \\ \frac{\partial N_i}{\partial y} \\ \frac{\partial N_i}{\partial z} \end{Bmatrix} = \boldsymbol{J}^{-1} \begin{Bmatrix} \frac{\partial N_i}{\partial \xi} \\ \frac{\partial N_i}{\partial \eta} \\ \frac{\partial N_i}{\partial \zeta} \end{Bmatrix}$$

to calculate the partial derivative of the shape function with respect to x, y coordinates.

Finally the element stiffness is

$$[K]^e = \iiint [B]^T[D][B]\mathrm{d}x\mathrm{d}y\mathrm{d}z = \int_{-1}^{1}\int_{-1}^{1}\int_{-1}^{1} [B]^T[D][B]|\boldsymbol{J}|\mathrm{d}\xi\mathrm{d}\eta\mathrm{d}\zeta \tag{10-51}$$

The element stiffness matrix contains ξ, η, ζ, coordinates, and the $[K]^e$ can be calculated by using the Gauss integration method, which will be studied in next chapter. The element stiffness can be expressed as

$$[K]^e = \begin{bmatrix} K_{11} & K_{12} & K_{13} & \cdots & K_{18} \\ K_{21} & K_{22} & K_{23} & \cdots & K_{28} \\ K_{31} & K_{32} & K_{33} & \cdots & K_{38} \\ \vdots & \vdots & \vdots & & \vdots \\ K_{81} & K_{82} & K_{83} & \cdots & K_{88} \end{bmatrix} \tag{10-52}$$

In Eq. (10−52), the element stiffness contains 576 (24×24) coefficients because each term in Eq. (10−52) contain nine coefficients.

10.7.3 Twenty-node 3D isoparametric element

Fig. 10−7 shows a distorted element and a standard twenty-node element, and the displacement functions for the standard element in the x-direction can be written as

$$\begin{aligned} u = {} & \alpha_1 + \alpha_2\xi + \alpha_3\eta + \alpha_4\zeta + \alpha_5\xi^2 + \alpha_6\eta^2 + \alpha_7\zeta^2 + \alpha_8\xi\eta + \alpha_9\eta\zeta + \\ & \alpha_{10}\xi\zeta + \alpha_{11}\xi^2\eta + \alpha_{12}\xi^2\zeta + \alpha_{13}\eta^2\xi + \alpha_{14}\eta^2\zeta + \alpha_{15}\zeta^2\xi + \\ & \alpha_{16}\zeta^2\eta + \alpha_{17}\xi\eta\zeta + \alpha_{18}\xi^2\eta\zeta + \alpha_{19}\xi\eta^2\zeta + \alpha_{20}\xi\eta\zeta^2 \end{aligned} \tag{10-53}$$

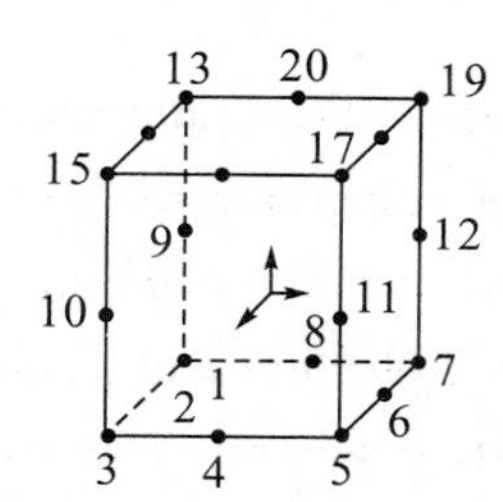

Fig. 10−13 Twenty-node element

Based on Fig. 10 − 13, the displacement can be expressed in terms of shape function as

$$\begin{cases} u = \sum\limits_{i=1}^{20} N_i(\xi,\eta,\zeta)u_i \\ v = \sum\limits_{i=1}^{20} N_i(\xi,\eta,\zeta)v_i \\ w = \sum\limits_{i=1}^{20} N_i(\xi,\eta,\zeta)w_i \end{cases} \tag{10-54}$$

where

$$
\begin{cases}
N_i = \frac{1}{8}(1+\xi_i\xi)(1+\eta_i\eta)(1+\zeta_i\zeta)(\xi_i\xi+\eta_i\eta+\zeta_i\zeta-2) \\
\qquad\qquad\qquad (i=1,\ 3,\ 5,\ 7,\ 13,\ 15,\ 17,\ 19) \\
N_i = \frac{1}{4}(1-\xi^2)(1+\eta_i\eta)(1+\zeta_i\zeta) \quad (i=2,\ 6,\ 14,\ 18) \\
N_i = \frac{1}{4}(1-\eta^2)(1+\xi_i\xi)(1+\zeta_i\zeta) \quad (i=4,\ 8,\ 16,\ 20) \\
N_i = \frac{1}{4}(1-\zeta^2)(1+\eta_i\eta)(1+\xi_i\xi) \quad (i=9,\ 10,\ 11,\ 12)
\end{cases}
\tag{10-55}
$$

Performing the same procedure as those for the eight-node 3D element in 10. 7. 2, the element stiffness can be written as

$$
[K]^e = \iiint [B]^T[D][B]\mathrm{d}x\mathrm{d}y\mathrm{d}z = \iiint [B]^T[D][B]|\boldsymbol{J}|\mathrm{d}\xi\mathrm{d}\eta\mathrm{d}\zeta \tag{10-56}
$$

The element stiffness matrix can be expressed as in Eq. (10−57). The element stiffness contains 3600 (60×60) coefficients because each term in Eq. (10−57) contains nine coefficients.

$$
[K]^e = \begin{bmatrix} K_{11} & K_{12} & K_{13} & \cdots & K_{120} \\ K_{21} & K_{22} & K_{23} & \cdots & K_{220} \\ K_{31} & K_{32} & K_{33} & \cdots & K_{320} \\ \vdots & \vdots & \vdots & & \vdots \\ K_{201} & K_{202} & K_{203} & \cdots & K_{2020} \end{bmatrix} \tag{10-57}
$$

Assignments:

1. What is the mainly advantage of isoparametric element in finite element analysis?

2. Under what conditions the isoparametric transformation cannot be performed? What should you pay attention to when making the mesh of isoparametric elements?

3. There is a distorted four-node element and a standard square element, as shown in the following figures. For the standard square element, the shape function is

$$
N_i = \frac{1}{4}(1-\xi)(1-\eta) \qquad N_j = \frac{1}{4}(1+\xi)(1-\eta)
$$

$$
N_k = \frac{1}{4}(1+\xi)(1+\eta) \qquad N_l = \frac{1}{4}(1-\xi)(1+\eta)
$$

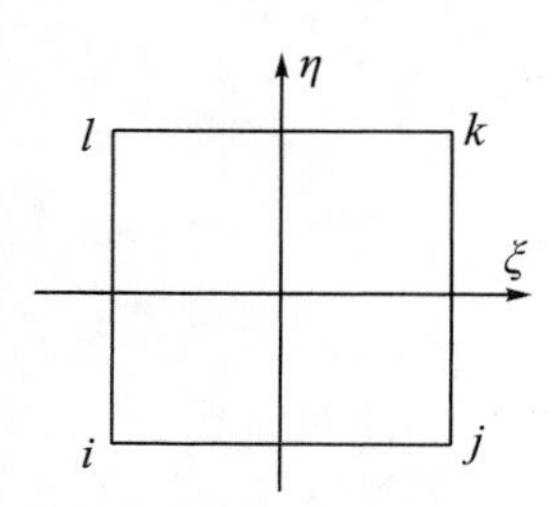

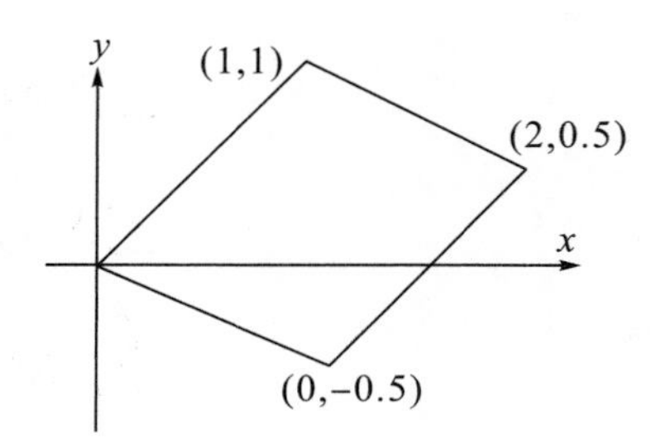

(1) Calculate the Jacobian matrix;

(2) Calculatethe derivative of shape function with respect to the x, y coordinates, i. e. $\begin{Bmatrix} \dfrac{\partial N_k}{\partial x} \\ \dfrac{\partial N_k}{\partial y} \end{Bmatrix}$.

4. The displacement functions of twenty-node standard element in x-direction is

$$u = \alpha_1 + \alpha_2\xi + \alpha_3\eta + \alpha_4\zeta + \alpha_5\xi^2 + \alpha_6\eta^2 + \alpha_7\zeta^2 + \alpha_8\xi\eta + \alpha_9\eta\zeta + \alpha_{10}\xi\zeta + \alpha_{11}\xi^2\eta + \alpha_{12}\xi^2\zeta + \alpha_{13}\eta^2\xi + \alpha_{14}\eta^2\zeta + \alpha_{15}\zeta^2\xi + \alpha_{16}\zeta^2\eta + \alpha_{17}\xi\eta\zeta + \alpha_{18}\xi^2\eta\zeta + \alpha_{19}\xi\eta^2\zeta + \alpha_{20}\xi\eta\zeta^2$$

If the displacement compatibility is satisfied in the mutual boundary between two adjacent elements?

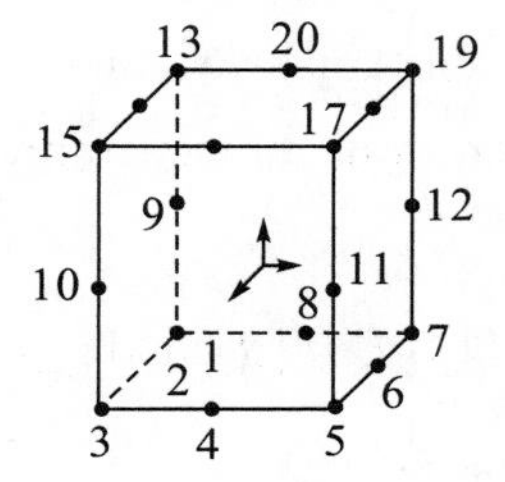

Chapter 11 Numerical integration

From the previous study, we can find that except for three-node triangle element and four-node tetrahedron element, for all the other type elements, such as six-node triangle element, rectangle element, ten-node tetrahedral element and brick element, the integrand in the element stiffness contains variables (x, y, or ξ, η), and we cannot move the integrand out of the integration. For example, the general 2D element stiffness can be written as

$$[K]^e = \int_{-1}^{1}\int_{-1}^{1}[B]^{\mathrm{T}}[D][B]t\,|\boldsymbol{J}|\,\mathrm{d}\xi\mathrm{d}\eta = \int_{-1}^{1}\int_{-1}^{1}[K'(\xi,\eta)]\mathrm{d}\xi\mathrm{d}\eta \tag{11-1}$$

where $[B]$ matrix contains variables ξ, η.

For such problem, some principles of numerical integration will be summarized in this chapter.

11.1 Newton-Cotes integration method

Supposing there is an integration $\int_a^b F(\xi)\mathrm{d}\xi$. We try to find a polynomial $\varphi(\xi)$ which at points i ($i=1, 2, \cdots, n$)

$$\varphi(\xi_i) = F(\xi_i) \tag{11-2}$$

then the integration can be approximated as

$$\int_a^b \varphi(\xi)\mathrm{d}\xi \approx \int_a^b F(\xi)\mathrm{d}\xi \tag{11-3}$$

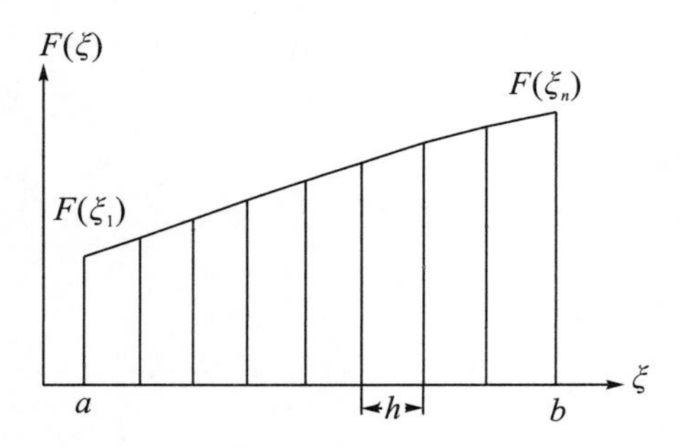

Fig. 11−1 A curve

In the follows, we will use Lagrange function $l_i^{n-1}(\xi)$ to express the function $\varphi(\xi)$, and $l_i^{n-1}(\xi)$ can be written as

$$l_i^{n-1}(\xi)=\prod_{k=1,k\neq i}^{n}\frac{\xi-\xi_k}{\xi_i-\xi_k}=\frac{(\xi-\xi_1)(\xi-\xi_2)\cdots(\xi-\xi_{i-1})(\xi-\xi_{i+1})\cdots(\xi-\xi_n)}{(\xi_i-\xi_1)(\xi_i-\xi_2)\cdots(\xi_i-\xi_{i-1})(\xi_i-\xi_{i+1})\cdots(\xi_i-\xi_n)} \tag{11-4}$$

where ξ must be one of the variables ξ_1, ξ_2, ξ_3, …, ξ_n. and

$$l_i^{n-1}(\xi_j)=\prod_{k=1,k\neq i}^{n}\frac{\xi_j-\xi_k}{\xi_i-\xi_k}=\frac{(\xi_j-\xi_1)(\xi_j-\xi_2)\cdots(\xi_j-\xi_{i-1})(\xi_j-\xi_{i+1})\cdots(\xi_j-\xi_n)}{(\xi_i-\xi_1)(\xi_i-\xi_2)\cdots(\xi_i-\xi_{i-1})(\xi_i-\xi_{i+1})\cdots(\xi_i-\xi_n)} \tag{11-5}$$

The function $l_i^{n-1}(\xi_j)$ has the property

$$l_i^{n-1}(\xi_j)=\prod_{k=1,k\neq i}^{n}\frac{\xi_j-\xi_k}{\xi_i-\xi_k}=\delta_{ij}=\begin{cases}1, & \text{as } i=j\\ 0, & \text{as } i\neq j\end{cases} \tag{11-6}$$

As $i=j$, the numerator and denominator in Eq. (11−5) are exactly same, so it equals 1.0. As $i\neq j$, because ξ_j must be one of the variables ξ_1, ξ_2, ξ_3, …, ξ_n, then in the numerator, there must have the case that $(\xi_j-\xi_j)$, and therefore, it equals 0.

If the distances between any two adjacent points are the same, this integration method is called Newton-Cotes integration method. The polynomial function can be expressed in terms of Lagrange function as

$$\varphi(\xi)=\sum_{i=1}^{n}l_i^{n-1}(\xi)F(\xi_i) \tag{11-7}$$

From Eq. (11−7), one can find at the point ξ_i, $\varphi(\xi_i)=F(\xi_i)$. Therefore, we have

$$\int_a^b\varphi(\xi)\mathrm{d}\xi=\int_a^b\sum_{i=1}^{n}l_i^{n-1}(\xi)F(\xi_i)\mathrm{d}\xi=\sum_{i=1}^{n}\left[\int_a^b l_i^{n-1}(\xi)\mathrm{d}\xi\right]F(\xi_i)=\sum_{i=1}^{n}H_iF(\xi_i) \tag{11-8}$$

where

$$H_i=\int_a^b l_i^{n-1}(\xi)\mathrm{d}\xi \tag{11-9}$$

It can be seen that H_i is not related to $F(\xi_i)$, and then we have

$$\int_a^b F(\xi)\mathrm{d}\xi\approx\int_a^b\varphi(\xi)\mathrm{d}\xi=\sum_{i=1}^{n}H_iF(\xi_i)+R_{n-1} \tag{11-10}$$

where R_i is the correction factor, and $F(\xi_i)$ is the value at the point ξ_i.

For example, if we choose two point integration, i. e. $n=2$ and $\xi_1=a$, $\xi_2=b$, and then we have

$$l_1^1=\frac{\xi-\xi_2}{\xi_1-\xi_2}=\frac{\xi-b}{a-b},\qquad l_2^1=\frac{\xi-\xi_1}{\xi_2-\xi_1}=\frac{\xi-a}{b-a}$$

From Eq. (11−9), the coefficients H_1 and H_2 can be calculated

$$H_1=\int_a^b l_1^1(\xi)\mathrm{d}\xi=\int_a^b\frac{\xi-b}{a-b}\mathrm{d}\xi=0.5(b-a) \tag{11-11}$$

$$H_2=\int_a^b l_2^1(\xi)\mathrm{d}\xi=\int_a^b\frac{\xi-a}{b-a}\mathrm{d}\xi=0.5(b-a) \tag{11-12}$$

For example, calculate the integration $\int_0^3 (2^r - r)dr$ where the exact solution is $\left(\frac{1}{\ln 2}2^r - \frac{r^2}{2}\right)\Big|_0^3 = 5.599$.

Solution: if we choose two point integration, then $\xi_1 = 0$, $\xi_2 = 3$, as shown in Fig. 11−2.

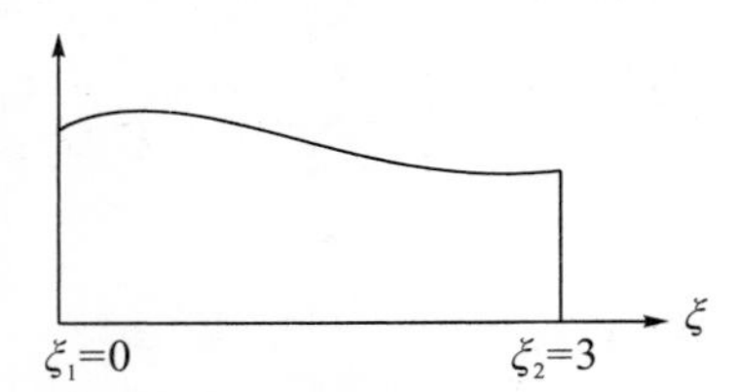

Fig. 11−2 Two point integration

The values at the tow points can be calculated

$$\begin{cases} F(\xi_1) = (2^\xi - \xi)|_{\xi=0} = 2^0 - 0 = 1 \\ F(\xi_2) = (2^\xi - \xi)|_{\xi=3} = 2^3 - 3 = 5 \end{cases} \tag{11−13}$$

From Eq. (11−9), we have

$$H_1 = H_2 = 0.5(b-a) = 0.5 \times (3-0) = 1.5 \tag{11−14}$$

Substituting Eq. (11−13) and Eq. (11−14) into Eq. (11−9), we can have

$$\int_0^3 (2^r - r)dr = \sum_{i=1}^{2} H_i F(\xi_i) = 1.5 \times 1 + 1.5 \times 5 = 9 \tag{11−15}$$

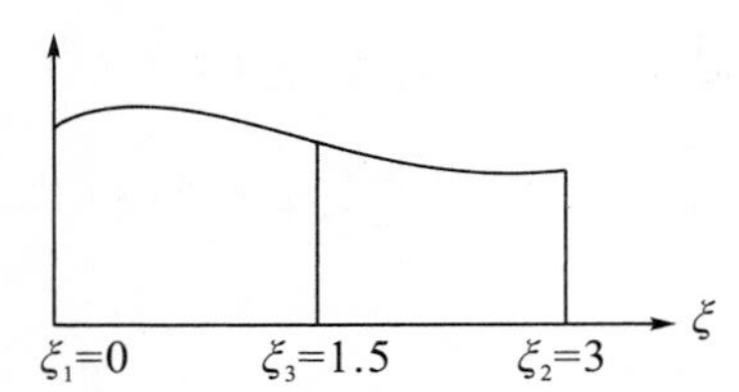

Fig. 11−3 Three-node integration

If we choose three point integration as shown in Fig. 11−3, then $\xi_1 = 0, \xi_2 = 1.5$, $\xi_3 = 3$. From Eq. (11−9), the coefficients can be obtained

$$H_1 = 0.5,\ H_2 = 2,\ H_3 = 0.5 \tag{11−16}$$

The values at the points ξ_1, ξ_2 and ξ_3 are

$$\begin{cases} F(\xi_1) = (2^\xi - \xi)|_{\xi=0} = 2^0 - 0 = 1 \\ F(\xi_2) = (2^\xi - \xi)|_{\xi=1.5} = 2^{1.5} - 1.5 = 1.328 \\ F(\xi_3) = (2^\xi - \xi)|_{\xi=3} = 2^3 - 3 = 5 \end{cases} \tag{11−17}$$

Substituting Eq. (11−16) and Eq. (11−17) into Eq. (11−9), we can have

$$\int_0^3 (2^r - r)dr = \sum_{i=1}^{2} H_i F(\xi_i) = 0.5 \times 1 + 2 \times 1.328 + 0.5 \times 5 = 5.657 \tag{11−18}$$

It can be seen that when we use the three point integration, the result close to the precise value, that means if we use more point integration, the results is more reliable.

11.2 Gauss integration method

Similar to Newton-Cotes integration, if the points is determined by the following method, the integration method is Gauss integration method.

The integration points ξ_i can be determined by

$$\int_a^b \xi^m P(\xi)\mathrm{d}\xi = 0 \quad (m = 0, 1, 2, \cdots, n-1) \tag{11-19}$$

where $P(\xi)$ is a polynomial and can be expressed as

$$P(\xi) = (\xi - \xi_1)(\xi - \xi_2)\cdots(\xi - \xi_n) \tag{11-20}$$

The function $\varphi(\xi)$ can be expressed as

$$\varphi(\xi) = \sum_{i=1}^{n} l_i^{n-1}(\xi)F(\xi_i) + R_i \tag{11-21}$$

where R_i is the correction factor.

The integration of the function $F(\xi)$ can be approximately replaced by

$$\int_a^b F(\xi)\mathrm{d}\xi \approx \int_a^b \varphi(\xi)\mathrm{d}\xi \approx \sum_{i=1}^{n}\int_a^b l_i^{n-1}(\xi)F(\xi_i)\mathrm{d}\xi = \sum_{i=1}^{n} H_i F(\xi_i) + R_{n-1} \tag{11-22}$$

where the coefficient H_i can be obtained from Eq. (11—9). It can be seen that the major difference between Gauss integration and Newton-Cotes integration method are the determination of the integration points.

For the same example, i. e. calculating $\int_0^3 (2^r - r)\mathrm{d}r$, we use the Gauss method to determine the integration points ξ_i .

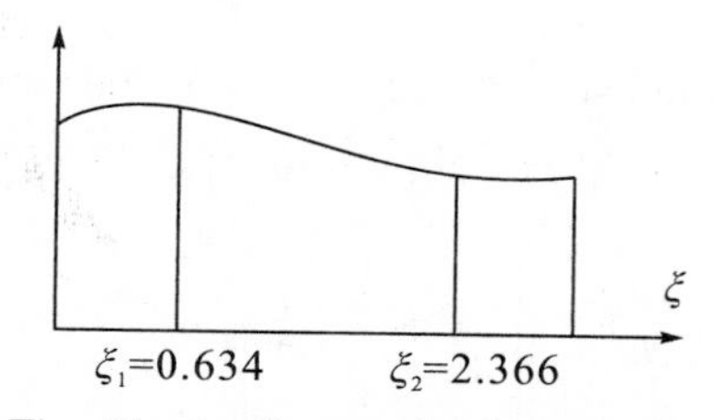

Fig. 11—4 Gauss point determination

When $m=0$, from Eq. (11—19), we have

$$\int_0^3 (\xi - \xi_1)(\xi - \xi_2)\mathrm{d}\xi = 0 \Rightarrow 3 - \frac{3}{2}(\xi_1 + \xi_2) + \xi_1\xi_2 = 0 \tag{11-23}$$

When $m=1$, we have

$$\int_0^3 (\xi - \xi_1)(\xi - \xi_2)\mathrm{d}\xi = 0 \Rightarrow \frac{9}{4} - (\xi_1 + \xi_2) + \frac{1}{2}\xi_1\xi_2 = 0 \tag{11-24}$$

Combing Eq. (11—23) with Eq. (11—24), we have

$$\begin{cases} \xi_1 + \xi_2 = 3 \\ \xi_1\xi_2 = \dfrac{3}{2} \end{cases} \tag{11-25}$$

Finally

$$\xi_1 = \frac{3-\sqrt{3}}{2} = 0.634, \quad \xi_2 = \frac{3+\sqrt{3}}{2} = 2.366 \tag{11-26}$$

From Eq. (11−9), i. e. $H_i = \int_a^b l_i^{n-1}(\xi)\mathrm{d}\xi$, the coefficients H_i can be obtained

$$\begin{cases} H_1 = \int_0^3 \frac{\xi - \xi_2}{\xi_1 - \xi_2}\mathrm{d}\xi = \frac{3}{2} \\ H_2 = \int_0^3 \frac{\xi - \xi_1}{\xi_1 - \xi_2}\mathrm{d}\xi = \frac{3}{2} \end{cases} \tag{11-27}$$

The values at the two integration points are

$$\begin{cases} F(\xi_1) = (2^\xi - \xi)\,|_{\xi=0.634} = 0.918 \\ F(\xi_2) = (2^\xi - \xi)\,|_{\xi=2.366} = 2.789 \end{cases} \tag{11-28}$$

Substituting Eqs. (11−27) and (11−28) into Eq. (11−22), we have

$$\int_0^3 (2^r - r)\mathrm{d}r = \sum_{i=1}^{2} H_i F(\xi_i) = H_1 F(\xi_1) + H_2 F(\xi_2) = 5.56 \tag{11-29}$$

It can be seen that the calculation result by the Gauss integration method is more precise than that by the Newton-Cotes integration method. Therefore, in the finite element analysis, Gauss integration method is adopted.

11.3 Gauss integration application in a standard element

11.3.1 Four point integration

First, we consider one dimensional integration in the range from −1 to +1. We take two point integration method ($n=2$) as shown in Fig. 11−5.

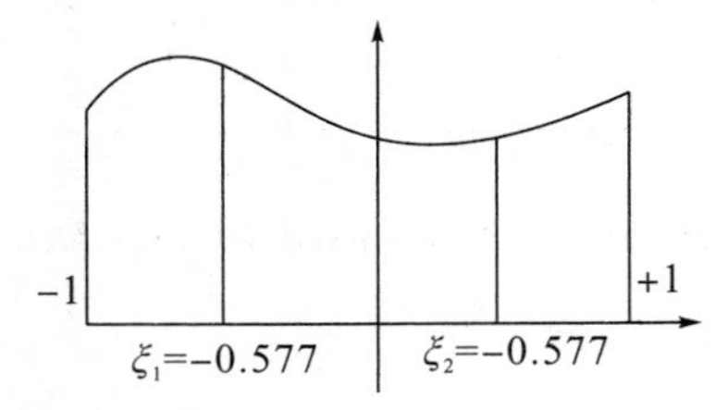

Fig. 11−5 Two Gauss points

When $m=0$, we have

$$\int_{-1}^{1} (\xi - \xi_1)(\xi - \xi_2)\mathrm{d}\xi = 0 \tag{11-30}$$

When $m=1$, we have

$$\int_{-1}^{1} \xi(\xi - \xi_1)(\xi - \xi_2)\mathrm{d}\xi = 0 \tag{11-31}$$

From Eqs. (11−30) and (11−31), we have

$$\xi_1 = -0.577, \ \xi_2 = 0.577$$

From Eq. (11−9), i. e. $H_i = \int_a^b l_i^{n-1}(\xi)\mathrm{d}\xi$, the coefficients H_i can be obtained

$$H_1 = \int_{-1}^{1} \frac{\xi - \xi_2}{\xi_1 - \xi_2}\mathrm{d}\xi = \int_{-1}^{1} \frac{\xi - 0.577}{-0.577 - 0.577}\mathrm{d}\xi = 1 \tag{11-32}$$

$$H_2 = \int_{-1}^{+1} \frac{\xi - \xi_1}{\xi_2 - \xi_1}\mathrm{d}\xi = \int_{-1}^{+} \frac{\xi + 0.577}{0.577 + 0.577}\mathrm{d}\xi = 1 \tag{11-33}$$

The element stiffness for the 2D case shown in Eq. (11−1) can be rewritten as

$$\begin{aligned}
[K]^e &= \int_{-1}^{1}\int_{-1}^{1} [B]^{\mathrm{T}}[D][B]\,t\,|\boldsymbol{J}|\,\mathrm{d}\xi\mathrm{d}\eta \\
&= \int_{-1}^{1}\int_{-1}^{1} [K'(\xi,\eta)]\mathrm{d}\xi\mathrm{d}\eta \\
&= \int_{-1}^{1}\left\{\int_{-1}^{1} [K'(\xi,\eta)]\mathrm{d}\xi\right\}\mathrm{d}\eta \\
&= \int_{-1}^{1}\left\{\sum_{j=1}^{n} H_j [K'(\xi_j,\eta)]\right\}\mathrm{d}\eta \\
&= \sum_{i=1}^{n} H_i \sum_{j=1}^{n} H_j [K'(\xi_j,\eta_i)] \\
&= \sum_{i=1}^{n}\sum_{j=1}^{n} H_i H_j [K'(\xi_j,\eta_i)]
\end{aligned} \tag{11-34}$$

where ξ and η are the horizontal and vertical axis.

In Eq. (11−34), the integrand $[K'(\xi,\eta)] = [B]^{\mathrm{T}}[D][B]t|\boldsymbol{J}|$. The $[B]$ matrix for four-node element studied in 10. 2 is

$$[B] = \begin{bmatrix}
\frac{\partial N_i}{\partial x} & 0 & \frac{\partial N_j}{\partial x} & 0 & \frac{\partial N_k}{\partial x} & 0 & \frac{\partial N_l}{\partial x} & 0 \\
0 & \frac{\partial N_i}{\partial y} & 0 & \frac{\partial N_j}{\partial y} & 0 & \frac{\partial N_k}{\partial y} & 0 & \frac{\partial N_l}{\partial y} \\
\frac{\partial N_i}{\partial y} & \frac{\partial N_i}{\partial x} & \frac{\partial N_j}{\partial y} & \frac{\partial N_j}{\partial x} & \frac{\partial N_k}{\partial y} & \frac{\partial N_k}{\partial x} & \frac{\partial N_l}{\partial y} & \frac{\partial N_l}{\partial x}
\end{bmatrix} \tag{11-35}$$

Therefore, $K'(\xi,\eta)$ is 8×8 matrix which can be expressed as

$$K'(\xi,\eta) = \sum_{i=1}^{n}\sum_{j=1}^{n} H_i H_j \begin{bmatrix}
k'_{11}(\xi,\eta) & k'_{12}(\xi,\eta) & k'_{13}(\xi,\eta) & \cdots & k'_{18}(\xi,\eta) \\
k'_{21}(\xi,\eta) & k'_{22}(\xi,\eta) & k'_{23}(\xi,\eta) & \cdots & k'_{28}(\xi,\eta) \\
k'_{31}(\xi,\eta) & k'_{32}(\xi,\eta) & k'_{33}(\xi,\eta) & \cdots & k'_{38}(\xi,\eta) \\
\vdots & \vdots & \vdots & & \vdots \\
k'_{81}(\xi,\eta) & k'_{82}(\xi,\eta) & k'_{83}(\xi,\eta) & \cdots & k'_{88}(\xi,\eta)
\end{bmatrix} \tag{11-36}$$

For a 2D standard square element ($n=2$), the four integration points are shown in Fig. 11−6.

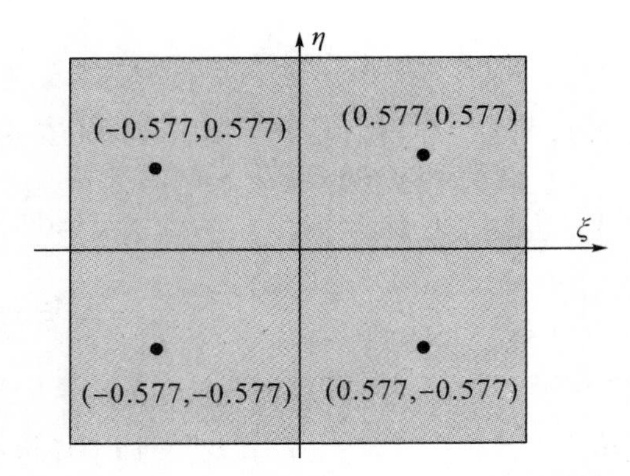

Fig. 11−6 Four integration points in a standard square element

The terms inside the matrix could be considered as little chicks inside eggs. If the little chicks want to come out, they have to break the eggshell. If the term $K'_{21}(\xi,\eta)$, for instance, becomes $K_{21}(\xi,\eta)$, it has to go through the following procedure

$$K_{21} = K'_{21}(0.577,0.577) + K'_{21}(-0.577,0.577) + K'_{21}(-0.577,-0.577) + K'_{21}(0.577,-0.577) \quad (11-37)$$

It can be seen that $K_{21}(\xi,\eta)$ is the sum of the $K'_{21}(\xi,\eta)$ at the four points.

11.3.2 Nine point integration

If we take three points for 1D integration, as shown in Fig. 11−7, according to Eq. (11−19), one can have

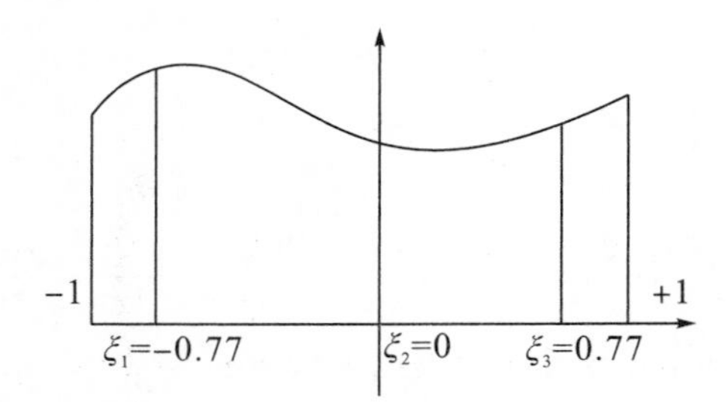

Fig. 11−7 Three gauss points

When $m=0$,

$$\int_{-1}^{1}(\xi-\xi_1)(\xi-\xi_2)\mathrm{d}\xi = 0 \quad (11-38)$$

When $m=1$,

$$\int_{-1}^{1}\xi(\xi-\xi_1)(\xi-\xi_2)\mathrm{d}\xi = 0 \quad (11-39)$$

When $m=2$,

$$\int_{-1}^{1}\xi^2(\xi-\xi_1)(\xi-\xi_2)(\xi-\xi_3)\mathrm{d}\xi = 0 \quad (11-40)$$

From Eqs. (11−38), (11−39) and (11−40), one can have

$$\xi_1 = -0.77, \quad \xi_2 = 0, \quad \xi_3 = 0.77$$

From Eq. (11−19), i. e. $H_i = \int_a^b l_i^{n-1}(\xi)\mathrm{d}\xi$, the coefficients H_1, H_2 and H_3 can be obtained

$$H_1 = \int_{-1}^{1} \frac{(\xi - \xi_2)(\xi - \xi_3)}{(\xi_1 - \xi_2)(\xi_1 - \xi_3)}\mathrm{d}\xi = 0.55556 \tag{11-41}$$

$$H_2 = \int_{-1}^{1} \frac{(\xi - \xi_1)(\xi - \xi_3)}{(\xi_2 - \xi_1)(\xi_2 - \xi_3)}\mathrm{d}\xi = 0.88889 \tag{11-42}$$

$$H_3 = \int_{-1}^{1} \frac{(\xi - \xi_1)(\xi - \xi_2)}{(\xi_3 - \xi_1)(\xi_3 - \xi_2)}\mathrm{d}\xi = 0.55556 \tag{11-43}$$

For the 2D standard element, the nine points are shown in Fig. 11−8.

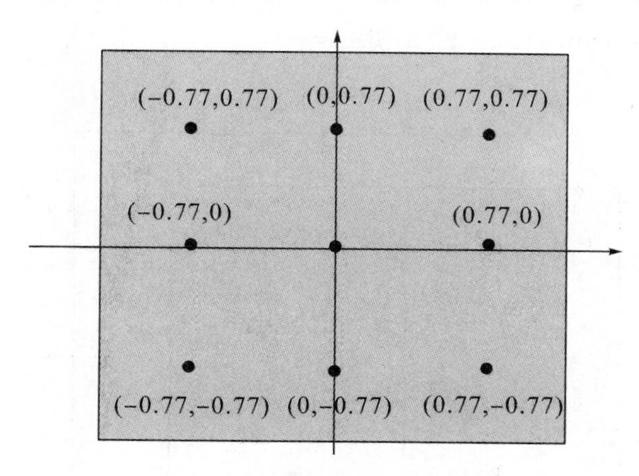

Fig. 11−8 Nine point integration

The element stiffness, i. e. $[K]^e = \int_{-1}^{1}\int_{-1}^{1}[B]^{\mathrm{T}}[D][B]t\,|\boldsymbol{J}|\,\mathrm{d}\xi\mathrm{d}\eta = \int_{-1}^{1}\int_{-1}^{1}[K']^e\mathrm{d}\xi\mathrm{d}\eta$, can be written as

$$K'(\xi,\eta) = \sum_{i=1}^{n}\sum_{j=1}^{n} H_i H_j \begin{bmatrix} k'_{11}(\xi,\eta) & k'_{12}(\xi,\eta) & k'_{13}(\xi,\eta) & \cdots & k'_{18}(\xi,\eta) \\ k'_{21}(\xi,\eta) & k'_{22}(\xi,\eta) & k'_{23}(\xi,\eta) & \cdots & k'_{28}(\xi,\eta) \\ k'_{31}(\xi,\eta) & k'_{32}(\xi,\eta) & k'_{33}(\xi,\eta) & \cdots & k'_{38}(\xi,\eta) \\ \vdots & \vdots & \vdots & & \vdots \\ k'_{81}(\xi,\eta) & k'_{82}(\xi,\eta) & k'_{83}(\xi,\eta) & \cdots & k'_{88}(\xi,\eta) \end{bmatrix} \tag{11-44}$$

Similarly, if the term $K'_{21}(\xi,\eta)$, for instance, becomes $K_{21}(\xi,\eta)$, it has to go through the following procedure

$$K_{21} = 0.556^2 k'_{21}(0.77, 0.77) + 0.556 \times 0.889 k'_{21}(0, 0.77) + \cdots + 0.556^2 k'_{21}(-0.77, -0.77) \tag{11-45}$$

11.3.3 Three dimensional integration method

Similarly, for 3D cases, the number of integration points should be 8 or 27. The element stiffness matrix can be written as

$$[K]^e = \int_{-1}^{1}\int_{-1}^{1}\int_{-1}^{1}[B]^{\mathrm{T}}[D][B]|\boldsymbol{J}|\,\mathrm{d}\xi\mathrm{d}\eta\mathrm{d}\zeta = \int_{-1}^{1}\int_{-1}^{1}\int_{-1}^{1}[K'(\xi,\eta,\zeta)]\mathrm{d}\xi\mathrm{d}\eta\mathrm{d}\zeta \tag{11-46}$$

The element stiffness can be solved by the Gauss integration method as

$$[K]^e = \int_{-1}^{1}\int_{-1}^{1}\int_{-1}^{1}[K'(\xi,\eta,\zeta)]\mathrm{d}\xi\mathrm{d}\eta\mathrm{d}\zeta = \sum_{i=1}^{n}\sum_{j=1}^{n}\sum_{k=1}^{n}H_iH_jH_k[K'(\xi_i,\eta_j,\zeta_k)] \tag{11-47}$$

For the eight point integration method as shown in Fig. 11−9, the coordinates $\xi=\pm0.577$, $\eta=\pm0.577$ and $\zeta=\pm0.577$. The corresponding coefficients $H=1.0$. The total number of integration points is $2\times2\times2=8$.

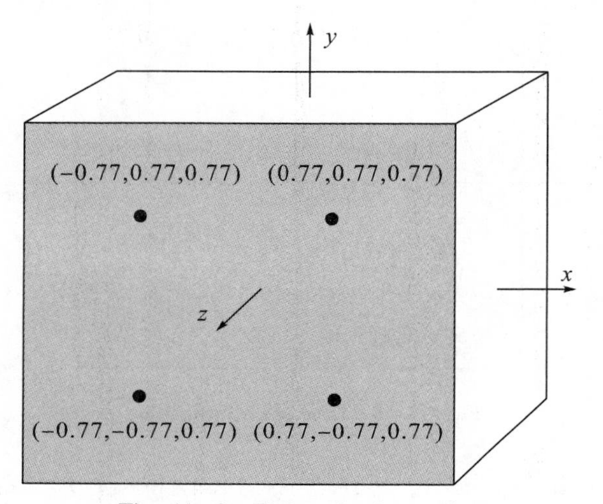

Fig. 11−9 Eight point integration

For the 27 point integration method as shown in Fig. 11−10, the coordinate $\xi=\pm0.577$, 0, $\eta=\pm0.577$, 0, $\zeta=\pm0.577$, 0. The corresponding coefficients $H(-0.577)=0.556$, $H(0)=0.889$ and $H(0.577)=0.556$.

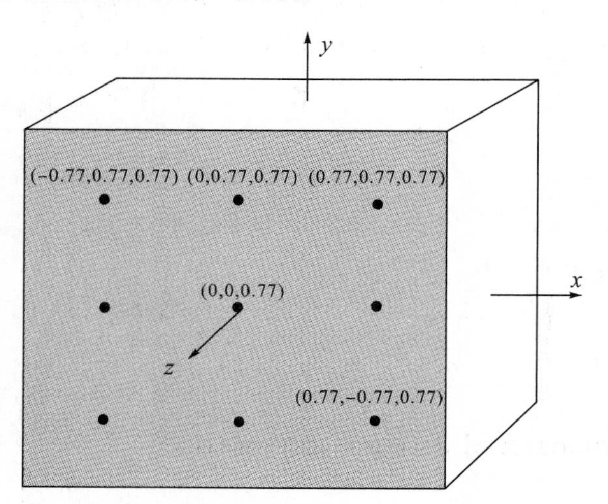

Fig. 11−10 Twenty-seven point integration

Then the total integration points will be $3\times3\times3=27$.

11.4 Equivalent nodal force

Four-node isoparametric element is selected to illustrate the calculation method of the equivalent node forces.

11.4.1 Concentrated load

As we know before, for the concentrated load as shown in Fig. 11-11, the equivalent nodal forces can be expressed as

$$\{F\}^e = \begin{Bmatrix} F_{1x} \\ F_{1y} \\ F_{2x} \\ F_{2y} \\ F_{3x} \\ F_{3y} \\ F_{4x} \\ F_{4y} \end{Bmatrix} = \begin{bmatrix} N_1 & 0 \\ 0 & N_1 \\ N_2 & 0 \\ 0 & N_2 \\ N_3 & 0 \\ 0 & N_3 \\ N_4 & 0 \\ 0 & N_4 \end{bmatrix} \begin{Bmatrix} P_x \\ P_y \end{Bmatrix} \tag{11-48}$$

where

$$N_1 = \frac{1}{4}(1-\zeta)(1-\eta)$$

$$N_2 = \frac{1}{4}(1+\zeta)(1-\eta)$$

$$N_3 = \frac{1}{4}(1+\zeta)(1+\eta)$$

$$N_4 = \frac{1}{4}(1-\zeta)(1+\eta)$$

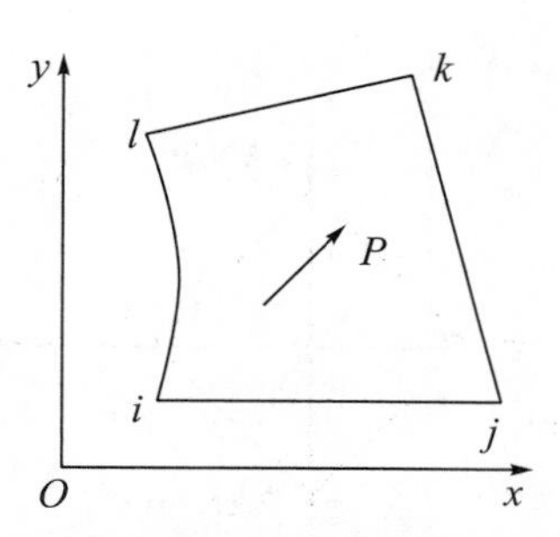

Fig. 11-11 A concentrated load acting on a point of a distorted element

11.4.2 Body load

For body forces, the equivalent nodal forces can be calculated by

$$\{F\}^e = \iint_A [N]^T \{p\} t \mathrm{d}x \mathrm{d}y = \int_{-1}^{1}\int_{-1}^{1} [N]^T \{p\} t |\boldsymbol{J}| \mathrm{d}\xi \mathrm{d}\eta \tag{11-49}$$

where

$$\{p\} = \begin{Bmatrix} 0 \\ \rho g \end{Bmatrix}$$

11.4.3 Distributed load

For distributed forces, the equivalent nodal forces can be calculated by

$$\{F\}^e = \int_L [N]^T \{p\} t \mathrm{d}s \tag{11-50}$$

where

$$\{p\} = \begin{Bmatrix} p_x \\ p_y \end{Bmatrix}$$

Supposing the distributed forces are acting along the edge $\xi = C$, then

$$\{F\}^e = \int_{-1}^{1} [N]^T \{p\} t s \mathrm{d}\eta \tag{11-51}$$

where

$$s = \sqrt{\left(\frac{\partial x}{\partial \eta}\right)^2 + \left(\frac{\partial y}{\partial \eta}\right)^2} \tag{11-52}$$

Assignments:

1. Under what conditions, you have to use Gaussian integration? How is the number of Gaussian points determined?

2. For a plane stress problem as shown in the following figure, the thickness of the plate $t = 1$ m, and the Poisson's ratio $\mu = 0$. Use four-node isoparametric element to calculate the displacements.

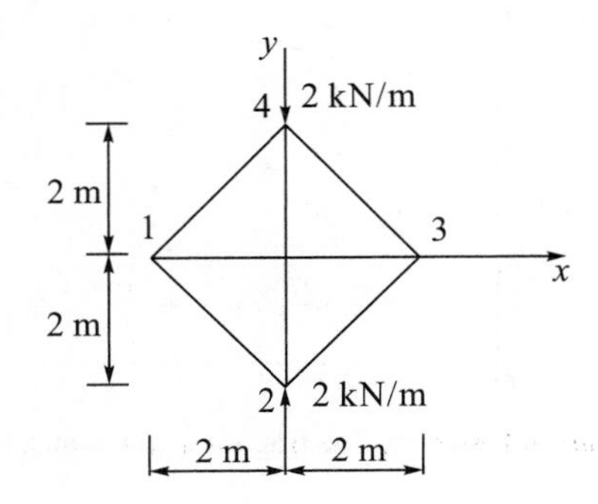

Chapter 12 Dynamic finite element method

In all the problems considered so far in this text, conditions that do not vary with time were generally assumed. There is little difficulty in extending the finite element idealization to situations that are time dependent.

The range of practical problems in which the time dimension has to be considered is great. Transient heat conduction, wave transmission in fluids and dynamic behavior of structures are typical examples. When displacements of an elastic body vary with time two sets of additional forces are called into play. The first is the inertia force, which for an acceleration characterized by $\ddot{u}$ can be replaced by its static equivalent. The second force is the damping force caused by the (frictional) resistances opposing the motion. This may be due to microstructure movements, air resistance, etc. , and are often related in a non-linear way to the velocity $\dot{u}$. For simplicity of treatment, however, only a linear viscous-type resistance will be considered.

In the first part of this chapter we shall formulate, by a simple extension of the triangle element methods used in previous chapters, and then we will consider the inertia force and the damping force.

12. 1 Formulation of time dependent problem

For the time dependent issue, all the variables (stresses, strains and displacements) are varying with time. For the triangle element under the nodal dynamic loading, as shown in Fig. 12—1, the displacements are

$$\begin{cases} u(t) = \alpha_1 + \alpha_2 x(t) + \alpha_3 y(t) \\ v(t) = \alpha_4 + \alpha_5 x(t) + \alpha_6 y(t) \end{cases} \tag{12—1}$$

The displacements can be rewritten as

$$\begin{Bmatrix} u(t) \\ v(t) \end{Bmatrix} = \begin{bmatrix} N_i & 0 & N_j & 0 & N_k & 0 \\ 0 & N_i & 0 & N_j & 0 & N_k \end{bmatrix} \begin{Bmatrix} u_i(t) \\ v_i(t) \\ u_j(t) \\ v_j(t) \\ u_k(t) \\ v_k(t) \end{Bmatrix} = [N(t)]\{\delta(t)\}^e \quad (12-2)$$

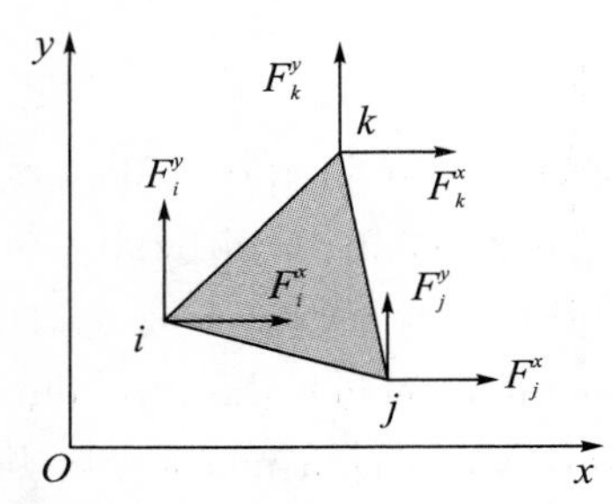

Fig. 12−1 A triangle element

Similarly, the strains can be expressed as

$$\{\varepsilon(t)\} = \begin{Bmatrix} \varepsilon_x(t) \\ \varepsilon_y(t) \\ \gamma_{xy}(t) \end{Bmatrix} = [B(t)]\{\delta(t)\}^e = \frac{1}{2A}\begin{bmatrix} b_i & 0 & b_j & 0 & b_k & 0 \\ 0 & c_i & 0 & c_j & 0 & c_k \\ c_i & b_i & c_j & b_j & c_k & b_k \end{bmatrix} \begin{Bmatrix} u_i(t) \\ v_i(t) \\ u_j(t) \\ v_j(t) \\ u_k(t) \\ v_k(t) \end{Bmatrix} \quad (12-3)$$

where $\begin{cases} b_i = y_j - y_k \\ c_i = -x_j + x_k \end{cases} \quad (i, j, k).$

The stresses can be expressed as $\{\sigma(t)\} = [S(t)]\{\delta(t)\}^e$, and it can be detailed as

$$\begin{Bmatrix} \sigma_x(t) \\ \sigma_y(t) \\ \tau_{xy}(t) \end{Bmatrix} = \frac{E}{2A(1-\nu^2)} \begin{bmatrix} b_i & \nu c_i & b_j & \nu c_j & b_k & \nu c_k \\ \nu b_i & c_i & \nu b_j & c_j & \nu b_k & c_k \\ \frac{1-\nu}{2}c_i & \frac{1-\nu}{2}b_i & \frac{1-\nu}{2}c_j & \frac{1-\nu}{2}b_j & \frac{1-\nu}{2}c_k & \frac{1-\nu}{2}b_k \end{bmatrix} \begin{bmatrix} u_i(t) \\ v_i(t) \\ u_j(t) \\ v_j(t) \\ u_k(t) \\ v_k(t) \end{bmatrix} \quad (12-4)$$

The stiffness can be expressed as

$$[K(t)]^e = \iint_A [B(t)]^{\mathrm{T}}[D][B(t)]T\,\mathrm{d}x\,\mathrm{d}y \quad (12-5)$$

where T is the thickness of the plate concerned.

Based on the principle of virtual displacement, one can have

$$\{F(t)\}^e = [K(t)]^e\{\delta(t)\}^e \quad (12-6)$$

More details of Eq. (12−6) can be expressed as

$$
\begin{bmatrix} F_i^x(t) \\ F_i^y(t) \\ F_j^x(t) \\ F_j^y(t) \\ F_k^x(t) \\ F_k^y(t) \end{bmatrix} = \frac{ET}{4(1-\nu^2)A}
\begin{bmatrix}
b_ib_i+\frac{1-\nu}{2}c_ic_i & \nu b_ic_i+\frac{1-\nu}{2}c_ib_i & b_ib_j+\frac{1-\nu}{2}c_ic_j & \nu b_ic_j+\frac{1-\nu}{2}c_ib_j & b_ib_k+\frac{1-\nu}{2}c_ic_k & \nu b_ic_k+\frac{1-\nu}{2}c_ib_k \\
\nu c_ib_i+\frac{1-\nu}{2}b_ic_i & c_ic_i+\frac{1-\nu}{2}b_ib_i & \nu c_ib_j+\frac{1-\nu}{2}b_ic_j & c_ic_j+\frac{1-\nu}{2}b_ib_j & \nu c_ib_k+\frac{1-\nu}{2}b_ic_k & c_ic_k+\frac{1-\nu}{2}b_ib_k \\
b_jb_i+\frac{1-\nu}{2}c_jc_i & \nu b_jc_i+\frac{1-\nu}{2}c_jb_i & b_jb_j+\frac{1-\nu}{2}c_jc_j & \nu b_jc_j+\frac{1-\nu}{2}c_jb_j & b_jb_k+\frac{1-\nu}{2}c_jc_k & \nu b_jc_k+\frac{1-\nu}{2}c_jb_k \\
\nu c_jb_i+\frac{1-\nu}{2}b_jc_i & c_jc_i+\frac{1-\nu}{2}b_jb_i & \nu c_jb_j+\frac{1-\nu}{2}b_jc_j & c_jc_j+\frac{1-\nu}{2}b_jb_j & \nu c_jb_k+\frac{1-\nu}{2}b_jc_k & c_jc_k+\frac{1-\nu}{2}b_jb_k \\
b_kb_i+\frac{1-\nu}{2}c_kc_i & \nu b_kc_i+\frac{1-\nu}{2}c_kb_i & b_kb_j+\frac{1-\nu}{2}c_kc_j & \nu b_kc_j+\frac{1-\nu}{2}c_kb_j & b_kb_k+\frac{1-\nu}{2}c_kc_k & \nu b_kc_k+\frac{1-\nu}{2}c_kb_k \\
\nu c_kb_i+\frac{1-\nu}{2}b_kc_i & c_kc_i+\frac{1-\nu}{2}b_kb_i & \nu c_kb_j+\frac{1-\nu}{2}b_kc_j & c_kc_j+\frac{1-\nu}{2}b_kb_j & \nu c_kb_k+\frac{1-\nu}{2}b_kc_k & c_kc_k+\frac{1-\nu}{2}b_kb_k
\end{bmatrix}
\begin{bmatrix} u_i(t) \\ v_i(t) \\ u_j(t) \\ v_j(t) \\ u_k(t) \\ v_k(t) \end{bmatrix}
\tag{12-7}
$$

For a structure with three elements and five nodes as shown in Fig. 12−2, the three element stiffness can be expressed as

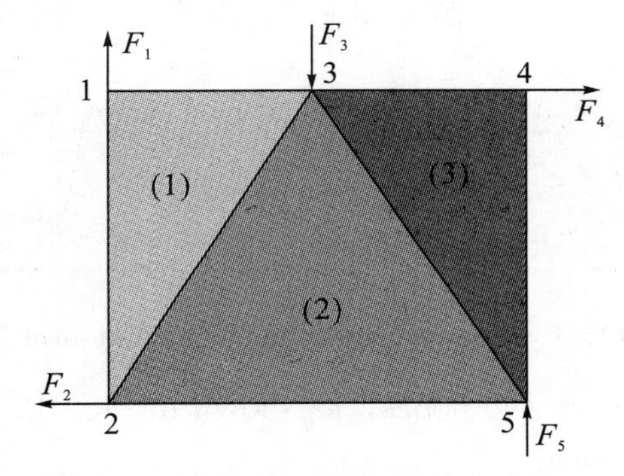

Fig. 12−2 Three elements and five nodes

$$
\begin{bmatrix} K_{22} & K_{23} & K_{21} \\ K_{32} & K_{33} & K_{31} \\ K_{12} & K_{13} & K_{11} \end{bmatrix}
\begin{Bmatrix} \{\delta_2\} \\ \{\delta_3\} \\ \{\delta_1\} \end{Bmatrix} =
\begin{Bmatrix} \{F_2\} \\ \{F_3\} \\ \{F_1\} \end{Bmatrix}
\tag{12-8}
$$

$$
\begin{bmatrix} K_{55} & K_{53} & K_{52} \\ K_{35} & K_{33} & K_{32} \\ K_{25} & K_{23} & K_{22} \end{bmatrix}
\begin{Bmatrix} \{\delta_5\} \\ \{\delta_3\} \\ \{\delta_2\} \end{Bmatrix} =
\begin{Bmatrix} \{F_5\} \\ \{F_3\} \\ \{F_2\} \end{Bmatrix}
\tag{12-9}
$$

$$
\begin{bmatrix} K_{55} & K_{54} & K_{53} \\ K_{45} & K_{44} & K_{43} \\ K_{35} & K_{34} & K_{33} \end{bmatrix}
\begin{Bmatrix} \{\delta_5\} \\ \{\delta_4\} \\ \{\delta_3\} \end{Bmatrix} =
\begin{Bmatrix} \{F_5\} \\ \{F_4\} \\ \{F_3\} \end{Bmatrix}
\tag{12-10}
$$

The global matrix can be expressed as

$$
\{F(t)\} = [K]\{\delta(t)\}
\tag{12-11}
$$

Eq. (12−11) can be detailed as

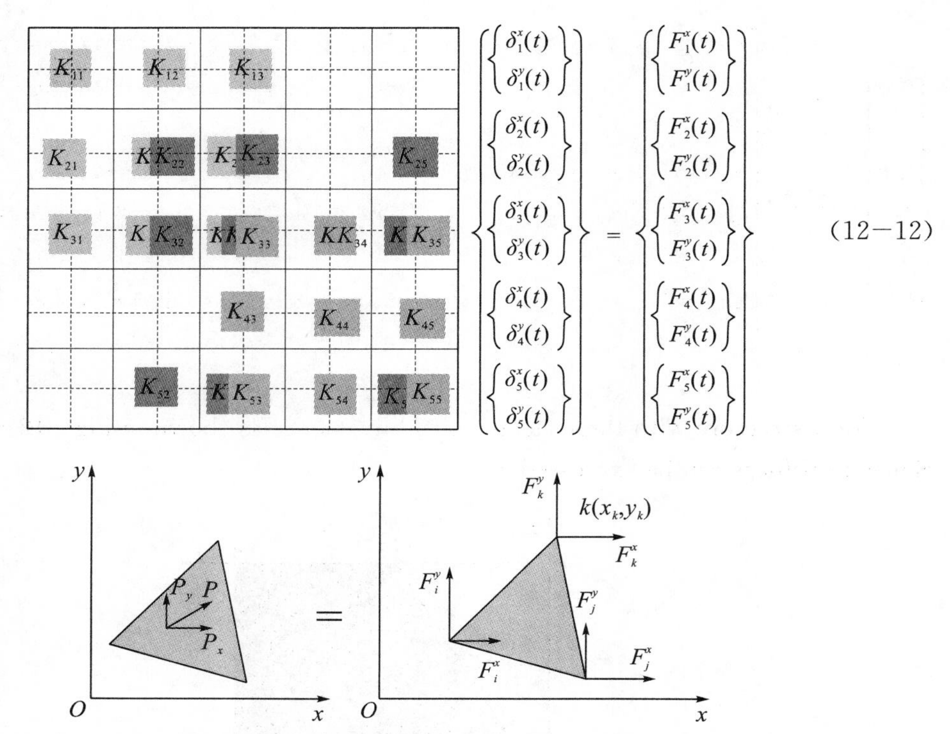

Fig. 12—3 A concentrated load and its equivalent forces

If a load is not just act on the nodes, as shown in Fig. 12—3, one can equivalently transfer it to the three nodes by

$$\{F(t)\}^e = [N]^{\mathrm{T}}\{P(t)\} \tag{12—13}$$

Eq. (12—13) can be detailed as

$$\{F(t)\}^e = \begin{Bmatrix} F_i^x(t) \\ F_i^y(t) \\ F_j^x(t) \\ F_j^y(t) \\ F_k^x(t) \\ F_k^y(t) \end{Bmatrix} = \begin{bmatrix} N_i & 0 \\ 0 & N_i \\ N_j & 0 \\ 0 & N_j \\ N_k & 0 \\ 0 & N_k \end{bmatrix} \begin{Bmatrix} P_x(t) \\ P_y(t) \end{Bmatrix} \tag{12—14}$$

where

$$N_i = \frac{1}{2A}(a_i + b_i x + c_i y) \quad (i,\ j,\ k) \tag{12—15}$$

$$\begin{cases} a_i = x_j y_k - x_k y_j \\ b_i = y_j - y_k \\ c_i = -x_j + x_k \end{cases} \quad (i,\ j,\ k)$$

For body force, as shown in Fig. 12—4, it also can be considered as a concentrated force acting on the gravity center (centroid).

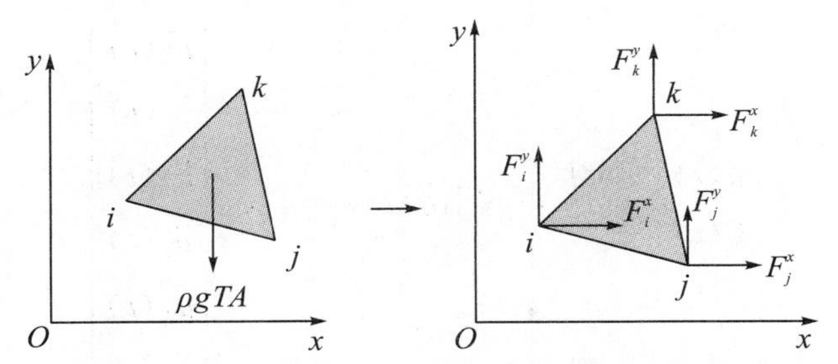

Fig. 12—4 The body force and equivalent nodal forces

According to the result of concentrate load, one can have

$$\{F\}^e = \begin{Bmatrix} F_i^x \\ F_i^y \\ F_j^x \\ F_j^y \\ F_k^x \\ F_k^y \end{Bmatrix} = \begin{bmatrix} N_i & 0 \\ 0 & N_i \\ N_j & 0 \\ 0 & N_j \\ N_k & 0 \\ 0 & N_k \end{bmatrix} \begin{Bmatrix} P_x \\ P_y \end{Bmatrix} = \begin{bmatrix} N_i & 0 \\ 0 & N_i \\ N_j & 0 \\ 0 & N_j \\ N_k & 0 \\ 0 & N_k \end{bmatrix} \begin{Bmatrix} 0 \\ -\rho gTA \end{Bmatrix} \tag{12—16}$$

where $N_i = N_j = N_k = \frac{1}{3}$. From Eq. (12—16), one can have

$$\{F\}^e = \begin{Bmatrix} F_i^x \\ F_i^y \\ F_j^x \\ F_j^y \\ F_k^x \\ F_k^y \end{Bmatrix} = \begin{bmatrix} N_i & 0 \\ 0 & N_i \\ N_j & 0 \\ 0 & N_j \\ N_k & 0 \\ 0 & N_k \end{bmatrix} \begin{Bmatrix} 0 \\ -\rho gTA \end{Bmatrix} = \begin{Bmatrix} 0 \\ -N_i\rho gTA \\ 0 \\ -N_j\rho gTA \\ 0 \\ -N_k\rho gTA \end{Bmatrix} = \begin{Bmatrix} 0 \\ -\frac{1}{3}\rho gTA \\ 0 \\ -\frac{1}{3}\rho gTA \\ 0 \\ -\frac{1}{3}\rho gTA \end{Bmatrix} \tag{12—17}$$

12.2 Inertial force

The inertial force is related to the acceleration $\ddot{u}$. Based on the relationship between the displacements of a point inside a triangle element and the three-node displacements, i. e.

$$\begin{Bmatrix} u(t) \\ v(t) \end{Bmatrix} = [N]\{\delta(t)\}^e \tag{12—18}$$

Taking the second derivative with respect to time, one can have

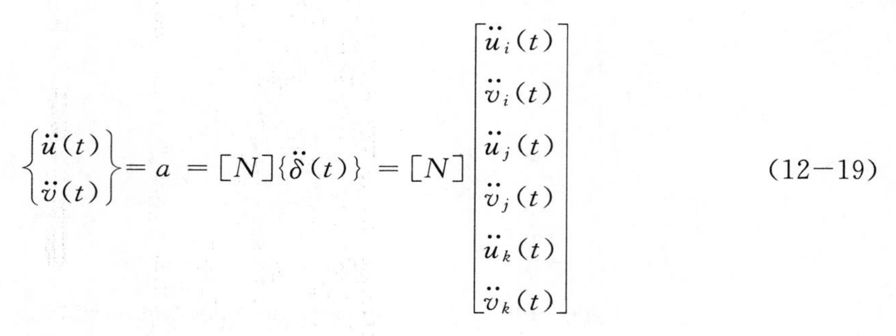

$$\begin{Bmatrix} \ddot{u}(t) \\ \ddot{v}(t) \end{Bmatrix} = a = [N]\{\ddot{\delta}(t)\} = [N]\begin{bmatrix} \ddot{u}_i(t) \\ \ddot{v}_i(t) \\ \ddot{u}_j(t) \\ \ddot{v}_j(t) \\ \ddot{u}_k(t) \\ \ddot{v}_k(t) \end{bmatrix} \tag{12-19}$$

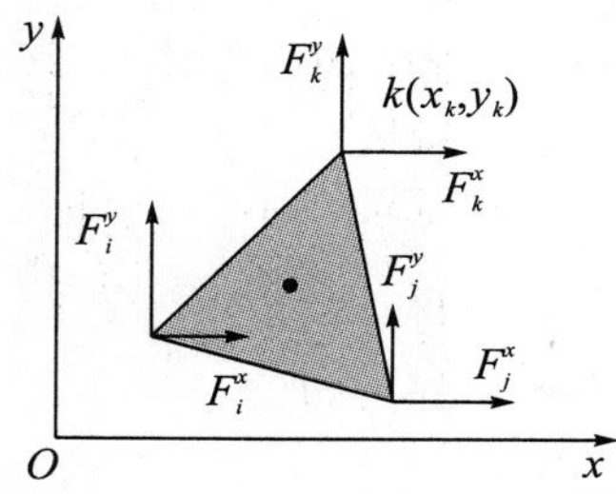

Fig. 12—5 Inertial force of an element

Suppose that the material density is ρ, then for a unit value the inertial force is $-\rho a$. We take a small square and the volume of the square is dV, then the equivalent nodal forces induced by this inertial force can be written as

$$\mathrm{d}\,\{F(t)\}_i^e = -[N]^{\mathrm{T}}\rho a\,\mathrm{d}V \tag{12-20}$$

then for the whole element, one can have

$$\{F(t)\}_i^e = -\iint [N]^{\mathrm{T}}\rho a\,\mathrm{d}V = -\iint \rho\,[N]^{\mathrm{T}}[N]\{\ddot{\delta}(t)\}^e T\,\mathrm{d}x\,\mathrm{d}y \tag{12-21}$$

where T is the thickness of the plate. Let

$$m = \iint \rho\,[N]^{\mathrm{T}}[N]\cdot T\,\mathrm{d}x\,\mathrm{d}y \tag{12-22}$$

Substituting Eq. (12—22) into Eq. (12—21), one can have

$$\{F(t)\}_i^e = -\iint \rho\,[N]^{\mathrm{T}}[N]T\,\mathrm{d}x\,\mathrm{d}y\{\ddot{\delta}(t)\}^e = -m\{\ddot{\delta}(t)\}^e \tag{12-23}$$

For triangle elements

$$[N] = \begin{bmatrix} N_i & 0 & N_j & 0 & N_k & 0 \\ 0 & N_i & 0 & N_j & 0 & N_k \end{bmatrix} \tag{12-24}$$

$$[N]^{\mathrm{T}} = \begin{bmatrix} N_i & 0 \\ 0 & N_i \\ N_j & 0 \\ 0 & N_j \\ N_k & 0 \\ 0 & N_k \end{bmatrix} \tag{12-25}$$

then, one can have

$$[N]^{\mathrm{T}}[N]=\begin{bmatrix} N_i & 0 \\ 0 & N_i \\ N_j & 0 \\ 0 & N_j \\ N_k & 0 \\ 0 & N_k \end{bmatrix}\begin{bmatrix} N_i & 0 & N_j & 0 & N_k & 0 \\ 0 & N_i & 0 & N_j & 0 & N_k \end{bmatrix}$$

$$=\begin{bmatrix} N_i^2 & 0 & N_iN_j & 0 & N_iN_k & 0 \\ 0 & N_i^2 & 0 & N_iN_j & 0 & N_iN_k \\ N_iN_j & 0 & N_j^2 & 0 & N_jN_k & 0 \\ 0 & N_iN_j & 0 & N_j^2 & 0 & N_jN_k \\ N_iN_k & 0 & N_jN_k & 0 & N_k^2 & 0 \\ 0 & N_iN_k & 0 & N_jN_k & 0 & N_k^2 \end{bmatrix} \tag{12-26}$$

There is one important formula which will be applied in solving Eq. (12—22), and it can be expressed as

$$\iint [N]_r\,[N]_s \mathrm{d}x\mathrm{d}y=\frac{A}{12}(1+\delta_{rs}) \tag{12-27}$$

where $\delta_{rs}=\begin{cases}1, \text{ when } r=s \\ 0, \text{ when } r\neq s\end{cases}$.

Combining with Eq. (12—26) and Eq. (12—27), Eq. (12—22) can be rewritten as

$$m=\iint \rho\,[N]^{\mathrm{T}}[N]\cdot T\mathrm{d}x\mathrm{d}y=\frac{\rho TA}{6}\begin{bmatrix} 1 & 0 & \frac{1}{2} & 0 & \frac{1}{2} & 0 \\ 0 & 1 & 0 & \frac{1}{2} & 0 & \frac{1}{2} \\ \frac{1}{2} & 0 & 1 & 0 & \frac{1}{2} & 0 \\ 0 & \frac{1}{2} & 0 & 1 & 0 & \frac{1}{2} \\ \frac{1}{2} & 0 & \frac{1}{2} & 0 & 1 & 0 \\ 0 & \frac{1}{2} & 0 & \frac{1}{2} & 0 & 1 \end{bmatrix} \tag{12-28}$$

So, the equivalent nodal forces caused by the inertial force of an element are

$$\{F(t)\}_i^e=\begin{Bmatrix} F_i^x(t) \\ F_i^y(t) \\ F_j^x(t) \\ F_j^y(t) \\ F_k^x(t) \\ F_k^y(t) \end{Bmatrix}=-\frac{\rho TA}{6}\begin{bmatrix} 1 & 0 & \frac{1}{2} & 0 & \frac{1}{2} & 0 \\ 0 & 1 & 0 & \frac{1}{2} & 0 & \frac{1}{2} \\ \frac{1}{2} & 0 & 1 & 0 & \frac{1}{2} & 0 \\ 0 & \frac{1}{2} & 0 & 1 & 0 & \frac{1}{2} \\ \frac{1}{2} & 0 & \frac{1}{2} & 0 & 1 & 0 \\ 0 & \frac{1}{2} & 0 & \frac{1}{2} & 0 & 1 \end{bmatrix}\begin{bmatrix} \ddot{u}_i(t) \\ \ddot{v}_i(t) \\ \ddot{u}_j(t) \\ \ddot{v}_j(t) \\ \ddot{u}_k(t) \\ \ddot{v}_k(t) \end{bmatrix} \tag{12-29}$$

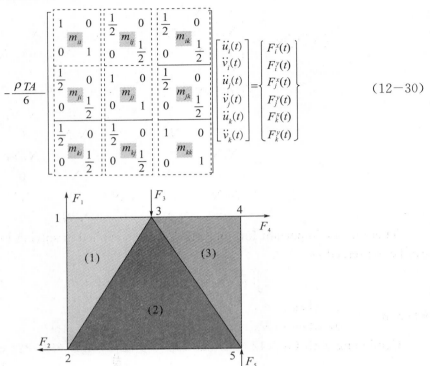

Eq. (12−29) can be rewritten as

$$-\frac{\rho TA}{6}\left[\begin{array}{c:c:c}\underbrace{\begin{matrix}1 & 0\\ 0 & 1\end{matrix}}_{m_{ii}} & \underbrace{\begin{matrix}\frac{1}{2} & 0\\ 0 & \frac{1}{2}\end{matrix}}_{m_{ij}} & \underbrace{\begin{matrix}\frac{1}{2} & 0\\ 0 & \frac{1}{2}\end{matrix}}_{m_{ik}}\\ \hdashline \underbrace{\begin{matrix}\frac{1}{2} & 0\\ 0 & \frac{1}{2}\end{matrix}}_{m_{ji}} & \underbrace{\begin{matrix}1 & 0\\ 0 & 1\end{matrix}}_{m_{jj}} & \underbrace{\begin{matrix}\frac{1}{2} & 0\\ 0 & \frac{1}{2}\end{matrix}}_{m_{jk}}\\ \hdashline \underbrace{\begin{matrix}\frac{1}{2} & 0\\ 0 & \frac{1}{2}\end{matrix}}_{m_{ki}} & \underbrace{\begin{matrix}\frac{1}{2} & 0\\ 0 & \frac{1}{2}\end{matrix}}_{m_{kj}} & \underbrace{\begin{matrix}1 & 0\\ 0 & 1\end{matrix}}_{m_{kk}}\end{array}\right]\begin{bmatrix}\ddot{u}_i(t)\\ \ddot{v}_i(t)\\ \ddot{u}_j(t)\\ \ddot{v}_j(t)\\ \ddot{u}_k(t)\\ \ddot{v}_k(t)\end{bmatrix}=\begin{Bmatrix}F_i^x(t)\\ F_i^y(t)\\ F_j^x(t)\\ F_j^y(t)\\ F_k^x(t)\\ F_k^y(t)\end{Bmatrix} \tag{12-30}$$

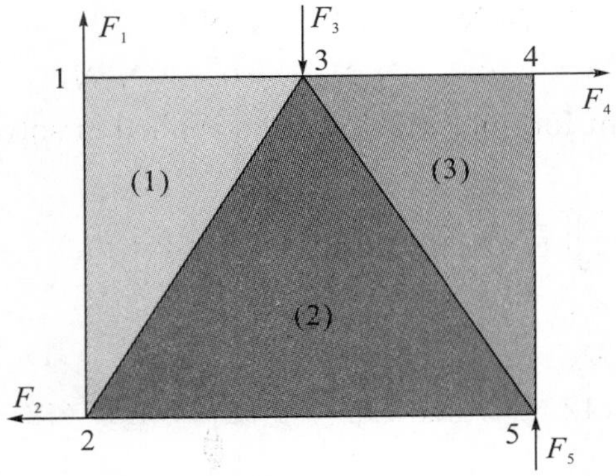

Fig. 12−6 Three elements and five nodes

For a structure with three elements, as shown in Fig. 12−6, the equilibrium equations of the three elements can be written as

$$\begin{bmatrix}m_{22} & m_{23} & m_{21}\\ m_{32} & m_{33} & m_{31}\\ m_{12} & m_{13} & m_{11}\end{bmatrix}\begin{Bmatrix}\{\ddot{\delta}_2\}\\ \{\ddot{\delta}_3\}\\ \{\ddot{\delta}_1\}\end{Bmatrix}=\begin{Bmatrix}\{F_2\}\\ \{F_3\}\\ \{F_1\}\end{Bmatrix}_i \tag{12-31}$$

$$\begin{bmatrix}m_{55} & m_{53} & m_{52}\\ m_{35} & m_{33} & m_{32}\\ m_{25} & m_{23} & m_{22}\end{bmatrix}\begin{Bmatrix}\{\ddot{\delta}_5\}\\ \{\ddot{\delta}_3\}\\ \{\ddot{\delta}_2\}\end{Bmatrix}=\begin{Bmatrix}\{F_5\}\\ \{F_3\}\\ \{F_2\}\end{Bmatrix}_i \tag{12-32}$$

$$\begin{bmatrix}m_{55} & m_{54} & m_{53}\\ m_{45} & m_{44} & m_{43}\\ m_{35} & m_{34} & m_{33}\end{bmatrix}\begin{Bmatrix}\{\ddot{\delta}_5\}\\ \{\ddot{\delta}_4\}\\ \{\ddot{\delta}_3\}\end{Bmatrix}=\begin{Bmatrix}\{F_5\}\\ \{F_4\}\\ \{F_3\}\end{Bmatrix}_i \tag{12-33}$$

For each element, $\{F(t)\}_i^e=-m\{\ddot{\delta}(t)\}^e$, for all the elements, one can have

$$\sum\{F(t)\}_i^e=-\sum m\{\ddot{\delta}(t)\}^e,\quad \text{i. e.}\quad \{F(t)\}_i=-M\{\ddot{\delta}(t)\} \tag{12-34}$$

where $M=\sum m$. Similar to the relationship between the global stiffness matrix and element stiffness matrix, the global M can be obtained as

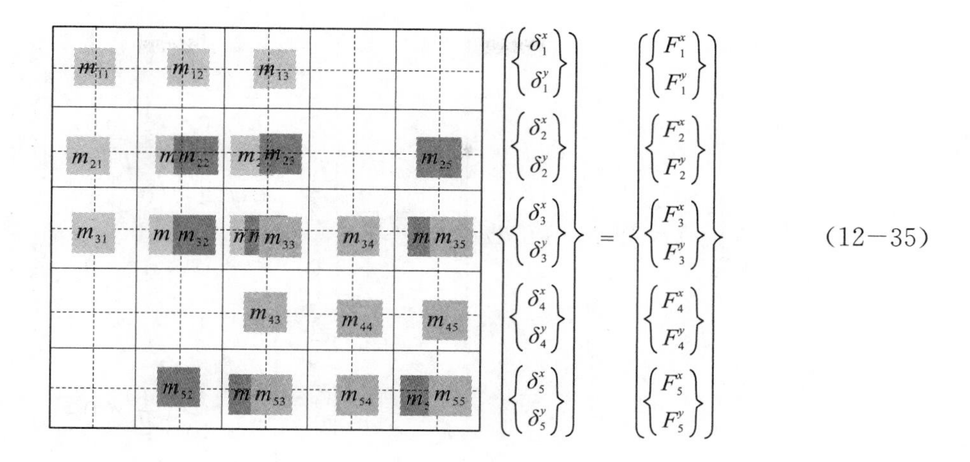

12.3 Damping force

The damping force is induced by the (frictional) resistances opposing the motion. It is often related in a non-linear way to the velocity $\dot{u}$. For simplicity, however, only a linear viscous-type resistance will be considered here.

Suppose the damping coefficient is γ, then for a unit value the damping force is $-\rho U$. Based on $\begin{Bmatrix} u(t) \\ v(t) \end{Bmatrix} = [N]\{\delta(t)\}^e$, the velocity U can be expressed as

$$U = \begin{Bmatrix} \dot{u}(t) \\ \dot{v}(t) \end{Bmatrix} = [N]\dot{\delta}(t) = [N]\begin{bmatrix} \dot{u}_i(t) \\ \dot{v}_i(t) \\ \dot{u}_j(t) \\ \dot{v}_j(t) \\ \dot{u}_k(t) \\ \dot{v}_k(t) \end{bmatrix} \tag{12-36}$$

The equivalent nodal forces are

$$\{F(t)\}_d^e = -\iint [N]^{\mathrm{T}} \gamma U \mathrm{d}V = -\iint \gamma [N]^{\mathrm{T}}[N]\{\dot{\delta}(t)\}^e T\mathrm{d}x\mathrm{d}y \tag{12-37}$$

where T is the thickness of the plate. Let

$$c = \iint \gamma [N]^{\mathrm{T}}[N] \cdot T\mathrm{d}x\mathrm{d}y \tag{12-38}$$

then, Eq. (12−37) can be rewritten as

$$\{F(t)\}_d^e = -\iint \gamma [N]^{\mathrm{T}}[N]\{\dot{\delta}(t)\}^e T\mathrm{d}x\mathrm{d}y = -c\{\dot{\delta}(t)\}^e \tag{12-39}$$

Comparing Eq. (12−38) with Eq. (12−22), one can find that the only difference is between the parameters of ρ and γ. Similar to Eq. (12−28), one can have

$$c=\iint\gamma\,[N]^{\mathrm{T}}[N]\cdot T\mathrm{d}x\mathrm{d}y=\frac{\gamma TA}{6}\begin{bmatrix}1&0&\frac{1}{2}&0&\frac{1}{2}&0\\0&1&0&\frac{1}{2}&0&\frac{1}{2}\\\frac{1}{2}&0&1&0&\frac{1}{2}&0\\0&\frac{1}{2}&0&1&0&\frac{1}{2}\\\frac{1}{2}&0&\frac{1}{2}&0&1&0\\0&\frac{1}{2}&0&\frac{1}{2}&0&1\end{bmatrix}\tag{12-40}$$

So, the equivalent nodal force caused by inertia are

$$\{F(t)\}_d^e=\begin{Bmatrix}F_i^x(t)\\F_i^y(t)\\F_j^x(t)\\F_j^y(t)\\F_k^x(t)\\F_k^y(t)\end{Bmatrix}=-\frac{\gamma TA}{6}\begin{bmatrix}1&0&\frac{1}{2}&0&\frac{1}{2}&0\\0&1&0&\frac{1}{2}&0&\frac{1}{2}\\\frac{1}{2}&0&1&0&\frac{1}{2}&0\\0&\frac{1}{2}&0&1&0&\frac{1}{2}\\\frac{1}{2}&0&\frac{1}{2}&0&1&0\\0&\frac{1}{2}&0&\frac{1}{2}&0&1\end{bmatrix}\begin{bmatrix}\dot{u}_i(t)\\\dot{v}_i(t)\\\dot{u}_j(t)\\\dot{v}_j(t)\\\dot{u}_k(t)\\\dot{v}_k(t)\end{bmatrix}\tag{12-41}$$

Eq. (12-41) can be rewritten as

$$-\frac{\gamma TA}{6}\left[\begin{array}{c:c:c}\underset{c_{ii}}{\begin{matrix}1&0\\0&1\end{matrix}}&\underset{c_{ij}}{\begin{matrix}\frac{1}{2}&0\\0&\frac{1}{2}\end{matrix}}&\underset{c_{ik}}{\begin{matrix}\frac{1}{2}&0\\0&\frac{1}{2}\end{matrix}}\\\hdashline\underset{c_{ji}}{\begin{matrix}\frac{1}{2}&0\\0&\frac{1}{2}\end{matrix}}&\underset{c_{jj}}{\begin{matrix}1&0\\0&1\end{matrix}}&\underset{c_{jk}}{\begin{matrix}\frac{1}{2}&0\\0&\frac{1}{2}\end{matrix}}\\\hdashline\underset{c_{ki}}{\begin{matrix}\frac{1}{2}&0\\0&\frac{1}{2}\end{matrix}}&\underset{c_{kj}}{\begin{matrix}\frac{1}{2}&0\\0&\frac{1}{2}\end{matrix}}&\underset{c_{kk}}{\begin{matrix}1&0\\0&1\end{matrix}}\end{array}\right]\begin{bmatrix}\dot{u}_i(t)\\\dot{v}_i(t)\\\dot{u}_j(t)\\\dot{v}_j(t)\\\dot{u}_k(t)\\\dot{v}_k(t)\end{bmatrix}=\begin{Bmatrix}F_i^x(t)\\F_i^y(t)\\F_j^x(t)\\F_j^y(t)\\F_k^x(t)\\F_k^y(t)\end{Bmatrix}\tag{12-42}$$

For the three element structure shown in Fig. 12-6, one can have

$$\{F(t)\}_d=-C\{\dot{\delta}(t)\}\tag{12-43}$$

where $C=\sum c$. Similar to Eq. (12-35), the global matrix in Eq. (12-43) can be detailed as

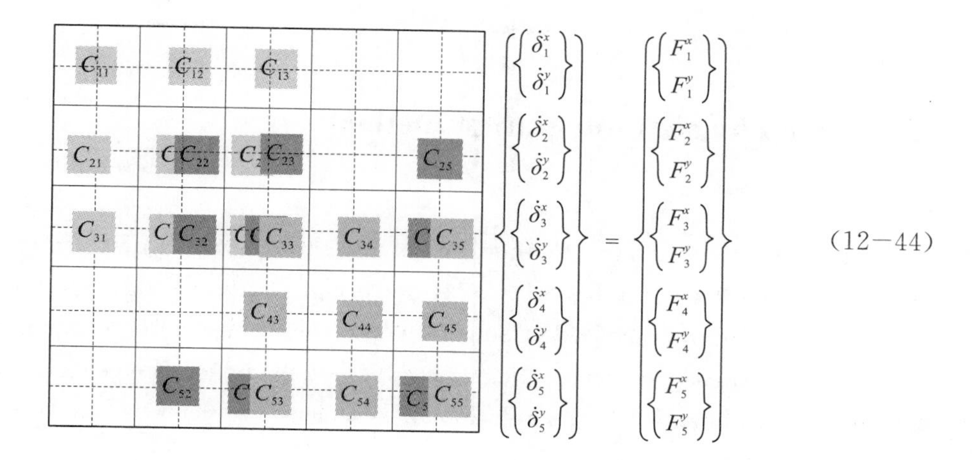

(12—44)

12.4 Global equilibrium equation

Except for the dynamic loads acting on the boundary of a structure, the inertial force and the damping force have to be considered, that means the total loads should be the sum of them, i. e.

$$\{F(t)\} = \{F(t)\}_l + \{F(t)\}_i + \{F(t)\}_d \tag{12—45}$$

where $\{F(t)\}_i = -M\{\ddot{\delta}(t)\}$, $\{F(t)\}_d = -C\{\dot{\delta}(t)\}$ and $\{F(t)\}_l$ is the extra loading acting on the boundary of the structure. Eq. (12—45) can be rewritten as

$$\{F(t)\} = \{F(t)\}_l - \{F(t)\}_i - \{F(t)\}_d = \{F(t)\}_l - M\ddot{\delta}(t) - C\dot{\delta}(t) \tag{12—46}$$

According to Eq. (12—11), one can have

$$\{F(t)\} = \{F(t)\}_l - M\ddot{\delta}(t) - C\dot{\delta}(t) = K\delta(t)$$

or

$$M\ddot{\delta}(t) + C\dot{\delta}(t) + K\delta(t) = \{F(t)\}_l \tag{12—47}$$

where

$$\begin{cases} K = \sum_{e=1}^{M} [K]^e \\ M = \sum_{e=1}^{M} m \\ C = \sum_{e=1}^{M} c \end{cases} \tag{12—48}$$

It can be seen that the global equilibrium equation, i. e. the ordinary differential equations, contains the first and second order derivatives. We cannot use the Gauss elimination method to solve the equation.

12.5 Step by step integration method

Generally, there are two methods in solving the ordinary differential equation of the characteristic form given in Eq. (12−47): one is eigenvalue method, and the second is the step by step integration method. Comparatively, the former is more complicated than the latter which is introduced here. The analytical solutions by using eigenvalue method are in general not economical for the solution of transient problems in linear cases and not applicable when non-linearity exists.

If the computation period is from 0 to the time t_0, we divide this period into n parts, i. e. the time step $\Delta t = t_0/n$. the differential equation in the time $t+\Delta t$ can be written as

$$M\ddot{\delta}_{t+\Delta t} + C\dot{\delta}_{t+\Delta t} + K\delta_{t+\Delta t} = \{F_{t+\Delta t}\}_l \tag{12−49}$$

where

$$\dot{\delta}_{t+\Delta t} = \dot{\delta}_t + \ddot{\delta}\Delta t \tag{12−50}$$

where $\ddot{\delta}$ is the acceleration in the period $[t, t+\Delta t]$, and generally

$$\ddot{\delta} = (1-\lambda)\ddot{\delta}_t + \lambda\ddot{\delta}_{t+\Delta t} \quad (0 \leqslant \lambda \leqslant 1) \tag{12−51}$$

Substituting Eq. (12−51) into Eq. (12−50), one can have

$$\dot{\delta}_{t+\Delta t} = \dot{\delta}_t + (1-\lambda)\ddot{\delta}_t \cdot \Delta t + \lambda\ddot{\delta}_{t+\Delta t} \cdot \Delta t \tag{12−52}$$

Based on Taylor's series, one can have

$$\delta_{t+\Delta t} = \delta_t + \dot{\delta}_t \cdot \Delta t + \frac{1}{2}\ddot{\delta} \cdot \Delta t^2 \tag{12−53}$$

Substituting Eq. (12−51) into Eq. (12−53), one can have

$$\delta_{t+\Delta t} = \delta_t + \dot{\delta}_t \cdot \Delta t + (1-\lambda)\ddot{\delta}_t \cdot \frac{\Delta t^2}{2} + \lambda\ddot{\delta}_{t+\Delta t} \cdot \frac{\Delta t^2}{2} \tag{12−54}$$

or

$$\ddot{\delta}_{t+\Delta t} = \frac{2}{\lambda\Delta t^2}(\delta_{t+\Delta t} - \delta_t) - \frac{2}{\lambda\Delta t}\dot{\delta}_t - \left(\frac{1}{\lambda} - 1\right)\ddot{\delta}_t \tag{12−55}$$

Substituting Eq. (12−55) into Eq. (12−52), one can have

$$\dot{\delta}_{t+\Delta t} = \frac{2}{\Delta t}(\delta_{t+\Delta t} - \delta_t) - \dot{\delta}_t \tag{12−56}$$

Substituting Eqs. (12−55) and (12−56) into Eq. (12−49), one can obtain

$$\left(K + \frac{2}{\lambda\Delta t^2}M + \frac{2}{\Delta t}C\right)\delta_{t+\Delta t} = F_{t+\Delta t} + M\left[\frac{2}{\lambda\Delta t^2}\delta_t + \frac{2}{\lambda\Delta t}\dot{\delta}_t + \left(\frac{1}{\lambda} - 1\right)\ddot{\delta}_t\right] + C\left(\frac{2}{\Delta t}\delta_t + \dot{\delta}_t\right) \tag{12−57}$$

where Δt is the time step, and λ is coefficient, and $0 \leqslant \lambda \leqslant 1$. From Eq. (12−57), we can obtain the displacements at the time $t+\Delta t$, that means all the displacements in various time can be obtained. From Eq. (12−56), the velocity can be calculated, and from

Eq. (12—55), the acceleration at the time $t+\Delta t$ can be obtained.

Assignments:

1. What is the major difference between dynamic finite element method and static finite element method?

2. Use four-node rectangle element to formulate the dynamic finite element method. Show the K, M and C matrix for the following structure, and show the global equilibrium equation.

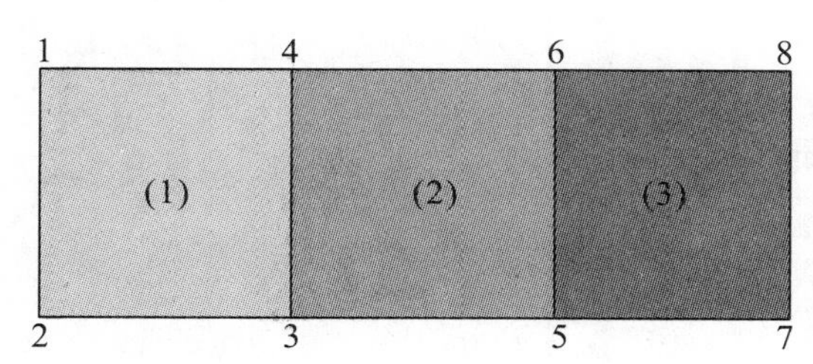

Chapter 13 Automatic Mesh Generation in MATLAB

13.1 Introduction

Mesh generation is widely applied in the pre-processes of Finite Element Method (FEM) because a structure could be generally divided into thousands of elements and it is very difficult to make the mesh manually. For example, a 2-D structure model consists of many triangle or quadrilateral elements, and a 3-D structure model can be similarly divided into a set of tetrahedron elements, hexahedron elements, and so on. In structural analysis, in order to obtain more precise results, the grid needs to be refined. Therefore, it is difficult to rely on manual mesh generation for solving problems of complex structures, and an efficient computer-aided mesh generation method has to be adopted. However, many mesh generators are often just used as "black boxes" and tend to be complex codes that are nearly inaccessible. In this chapter, we will introduce a method of automatic mesh generation in MTLAB. Compared with other programming software, we know that MALAB has strong features within matrix operations and visualization function. The purpose of this chapter is to introduce the contents of the MATLAB codes and how the MATLAB codes work in detail.

DistMesh is a simple MATLAB code for generation of unstructured triangular and tetrahedral meshes. It was developed by Per-Olof Persson (now at UC Berkeley) and Gilbert Strang in the Department of Mathematics at MIT. The code is short and simple. For the actual mesh generation, the iterative technique is based on the physical analogy between a simplex mesh and a truss structure. Mesh points are nodes of the truss. Assuming an appropriate force-displacement function for the bars in the truss at each iteration, and then we solve for equilibrium. The forces move the nodes, and the *Delaunay* triangulation algorithm adjusts the topology which determines the edges in every iteration.

The reference paper is: Per-Olof Persson and Gilbert Strang. DistMesh-A Simple Mesh Generator in MATLAB. SIAM Review, Volume 46 (2), pp. 329-345, June 2004. Website: http://persson.berkeley.edu/distmesh/.

13.2 The algorithm for mesh generation

In plane problem, the mesh generation algorithm is based on a simple mechanical analogy between a triangular mesh and a 2-D truss structure, or equivalently a structure of springs. Any set of points in the x, y-plane can be triangulated by the *Delaunay* algorithm. In the physical model, the edges of the triangles (the connections between pairs of points) correspond to bars, and the points correspond to joints of the truss. Each bar has a force-displacement relationship $f(l, l_0)$ depending on its current length l and its original or undeformed length l_0.

The external forces on the structure come at the boundaries. At every boundary node, there is a reaction force acting normal to the boundary. The magnitude of this force is precisely large enough to keep the node from moving outside. The positions of the joints are found by solving for a static force equilibrium in the structure. The hope is that the lengths of all the bars at equilibrium will be mostly equal when $h(x, y)=1$, so that we can finally get a well-shaped triangular mesh. To solve for the force equilibrium, the first step is collecting the x-coordinates and y-coordinates of all N mesh-points into an N-by-2 array p

$$p = [x \ y] \tag{13-1}$$

The force vector $F(p)$ has horizontal and vertical components at each meshpoint

$$F(p) = [F_{int,x}(p) \ F_{int,y}(p)] + [F_{ext,x}(p) \ F_{ext,y}(p)] \tag{13-2}$$

where F_{int} contains the internal forces from the bars, and F_{ext} are the external reaction forces from the boundaries. The first column of F contains the x-components of the forces, and the second column contains the y-components.

Note that $F(p)$ depends on the topology of the bars connecting the joints. In the algorithm, this structure is given by the *Delaunay* triangulation of the meshpoints. The *Delaunay* algorithm determines non-overlapping triangles that fill the convex hull of the input points, such that every edge is shared by at most two triangles, and the circumcircle of every triangle contains no other meshpoints. (Fig. 13−1)

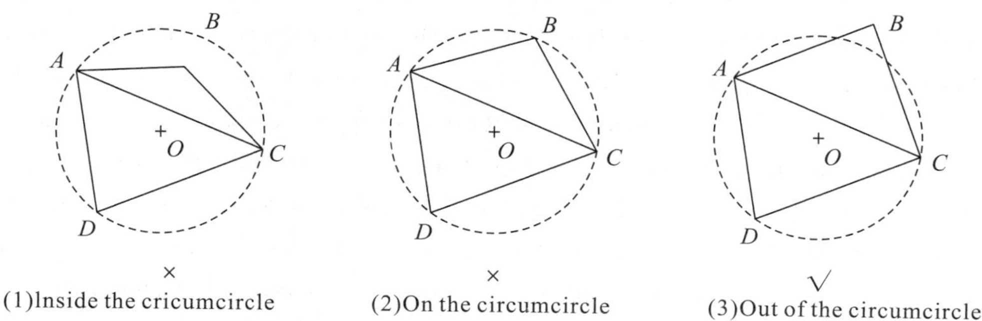

Fig. 13−1 The circumcircle of every triangle contains no other input points which means that all circumcircles shouldn't completely overlap each other

Note that the force vector $F(p)$ is not a continuous function of p, since the topology (the presence or absence of connecting bars) is changed by the *Delaunay* algorithm as the points move away from original locations.

The equation $F(p)=0$ has to be solved for a set of equilibrium positions p. This is a relatively hard problem, partly because of the discontinuity in the force function, and partly because of the external reaction forces at the boundaries. A simple approach to solve $F(p)=0$ is to introduce an artificial time-dependence

$$\frac{dp}{dt} = F(p),\ t \geqslant 0 \tag{13-3}$$

If a stationary solution is found, it satisfies our equation $F(p)=0$. The Eq. (13−3) is approximated using the forward Euler's method. At the discretized artificial time $t_n = n\Delta t$, the approximate solution $p_n \approx p(t_n)$ is updated by

$$p_{n+1} = p_n + \Delta t \cdot F(p_n) \tag{13-4}$$

The external reaction forces enter in the following way: All points that go outside the region during the update from p_n to p_{n+1} are moved back to the closest boundary locations. This conforms to the requirement that forces act normal to the boundary. The points can move along the boundary strictly just like wheels.

As for the force-displacement function $f(l,l_0)$ in each bar, the implementation uses this linear response for the repulsive forces, but it allows no attractive forces

$$f(l,l_0) = \begin{cases} k(l_0 - l), l < l_0 \\ 0, \qquad l \geqslant l_0 \end{cases} \tag{13-5}$$

where k is similar to stiffness coefficient and we set $k = 1$.

It is reasonable to require $f = 0$ for $l = l_0$. The proposed treatment of the boundaries means that no points are forced to stay at the boundary, and they are just prevented from crossing it. Therefore, it is important that the forces in most of the bars should be repulsive, to help the points spread out across the whole geometry. This means that $f(l,l_0)$ should be positive when the current length l is close to the desired length l_0, which can be achieved by choosing l_0 slightly larger than the length we actually desire (a good default in 2-D is 20%, which is expressed as *Fscale* = 1.2).

In the implementation, the desired edge length distribution is provided by the user as an element size function $h(x,y)$. Note that $h(x,y)$ does not have to equal the actual size, and alternatively it gives the relative distribution over the domain. For example, if $h(x,y) = 1 + x$ in the unit square, the edge lengths close to the left boundary ($x = 0$) will be about half of the edge lengths close to the right boundary ($x = 1$). This is true regardless of the number of points and the actual element sizes. To find the scaling, we need to compute the ratio between the actual edge lengths l_i of every bar and the desired size $h(x,y)$ at the midpoints (x_i, y_i) of every bar, and we regard the desired bars as the original or undeformed bars:

$$scaling\ factor = \left(\frac{\sum l_i^2}{\sum h(x_i, y_i)^2} \right)^{1/2} \tag{13-6}$$

Assuming here that $h(x,y)$ is specified by the user. It could also be created by using adaptive logic, such as the distance between the boundaries of the region (see Fig. 13—8 (5)), to implement the expression of local feature size. For highly curved boundaries, $h(x,y)$ could be expressed in terms of the curvature computed from $d(x,y)$. In addition, an adaptive solver can choose $h(x,y)$ to refine the mesh for good solutions.

The initial node positions p_0 can be defined in many ways. A random distribution of the points usually works well. For meshes intended to have uniform element sizes, good results are achieved by starting from equally spaced points. When a non-uniform size distribution $h(x,y)$ is desired, the convergence is faster if the initial distribution is weighted by probabilities proportional to $1/[h(x,y)]^2$. The rejection method starts with a uniform initial mesh inside the domain and discards points by using this probability.

13.3 Implementation

The complete source code for the two-dimensional mesh generator is shown as below. Each line is explained in detail.

The first line specifies the calling syntax for the function *distmesh2d*:

function [*p*, *t*] =*distmesh2d* (*fd*, *fh*, h_0, *box*, *pfix*, *varargin*)

This meshing function produces the following outputs:

• The node positions *p*. This *N*-by-2 array contains the *x*, *y* coordinates for each of the *N* nodes.

• The triangle indices *t*. Every row associated with each triangle has 3 integers to specify node numbers in a triangle.

The input arguments are as follows:

• The geometry is given as a distance function *fd*. This function returns the signed distance from each node location *p* to the closest point of the boundary.

• The relative desired edge length function $h(x,y)$ is given as a function *fh*, which returns an array *h* with one column for all meshpoints.

• The parameter h_0 is the distance between points in the initial distribution p_0. For uniform meshes ($h(x,y)$ = constant), the element size in the final mesh will usually be a little larger than this input.

• The bounding box for the region is an array $box = [x_{min}, y_{min}; x_{max}, y_{max}]$.

• The fixed node positions are given as an array *pfix* with two columns.

• Additional parameters to the functions *fd* and *fh* can be given in the last arguments *varargin* (type help *varargin* in MATLAB for more information).

At the beginning of the code, six parameters are set. The default values seem to work very generally, and for most purposes they can be left unmodified. The algorithm will stop when all movements in an iteration (relative to the average bar length) are smaller than the parameter *dptol*. Similarly, *ttol* controls how far the points can move relatively before a retriangulation by the *Delaunay* algorithm.

The 'internal pressure' is controlled by *Fscale*. The time step in Euler's method (13-4) is *deltat*, and *geps* is the tolerance in the geometry evaluations. The square root *eps* of the machine tolerance is the Δx and Δy in the numerical differentiation of the distance function. This is optimal for first order partial differential. The numbers *geps* and *deps* are scaled with the element size h_0, in case someone wants to mesh an atom or a galaxy in different units.

Distmesh2d Code

```
function [p,t] = distmesh2d (fd,fh,h0,box,pfix,varargin)
dptol = 0.001; ttol = 0.1; Fscale = 1.2; deltat = 0.2; geps = 0.001 * h0;
deps = sqrt(eps) * h0;

% 1. Create initial distribution in bounding box (equilateral triangles)
[x,y] = meshgrid(box(1,1):h0:box(2,1),box(1,2):h0 * sqrt(3)/2:box(2,2));
x(2:2:end,:) = x(2:2:end,:) + h0/2;        % Shift even rows
p = [x(:),y(:)];        % List of node coordinates

% 2. Remove points outside the region,apply the rejection method
p = p(feval(fd,p,varargin{:}) < geps,:);        % Keep only d < 0 points
r0 = 1./feval(fh,p,varargin{:}).^2;        % Probability to keep point
p = [pfix;p(rand(size(p,1),1) < r0./max(r0),:)];        % Rejection method
N = size(p,1);        % Number of points N
pold = inf;        % The old points before the last iteration,this is the first iteration
while 1

% 3. Retriangulation by the Delaunay algorithm
if max (sqrt(sum((p - pold).^2,2))/h0) > ttol        % Any large movement?
pold = p;        % Save current positions
t = delaunayn(p);        % List of triangles
pmid = (p(t(:,1),:) + p(t(:,2),:) + p(t(:,3),:))/3;        % Compute centroids
t = t(feval(fd,pmid,varargin{:}) <- geps,:);        % Keep interior triangles

% 4. Describe each bar by a unique pair of nodes
bars = [t(:,[1,2]);t(:,[1,3]);t(:,[2,3])];        % Interior bars duplicated
```

```
bars = unique(sort(bars,2),'rows');        % Bars as node pairs

% 5. Graphical output of the current mesh
trimesh(t,p(:,1),p(:,2),zeros(N,1))
view(2),axis equal,axis off,drawnow
end

% 6. Move mesh points based on bar lengths L and forces F
barvec = p(bars(:,1),:) - p(bars(:,2),:);        % List of bar vectors
L = sqrt(sum(barvec.^2,2));        % L = Bar lengths
hbars = feval(fh,(p(bars(:,1),:) + p(bars(:,2),:))/2,varargin{:});
% At the midpoints of the bars
L0 = hbars * Fscale * sqrt(sum(L.^2)/sum(hbars.^2));        % L0 = Desired lengths
F = max(L0 - L,0);        % Bar forces (scalars),k = 1
Fvec = F./L * [1,1].* barvec;        % Bar forces (x,y components)
Ftot = full(sparse(bars(:,[1,1,2,2]),ones(size(F)) * [1,2,1,2],[Fvec, - Fvec],N,2));
Ftot(1:size(pfix,1),:) = 0;        % Force = 0 at fixed points
p = p + deltat * Ftot;        % Update node positions

% 7. Bring outside points back to the boundary
d = feval(fd,p,varargin{:});   ix = d > 0;        % Find points outside (d > 0)
dgradx = (feval(fd,[p(ix,1) + deps,p(ix,2)],varargin{:}) - d(ix))/deps;
%Numerical
dgrady = (feval(fd,[p(ix,1),p(ix,2) + deps],varargin{:}) - d(ix))/deps;
% Numerical gradient
p(ix,:) = p(ix,:) - [d(ix).* dgradx,d(ix).* dgrady];
% Project back to boundary

% 8. Termination criterion: All interior nodes move less than dptol (scaled)
if max(sqrt(sum(deltat * Ftot(d <- geps,:).^2,2))/h0) < dptol,break; end
end
```

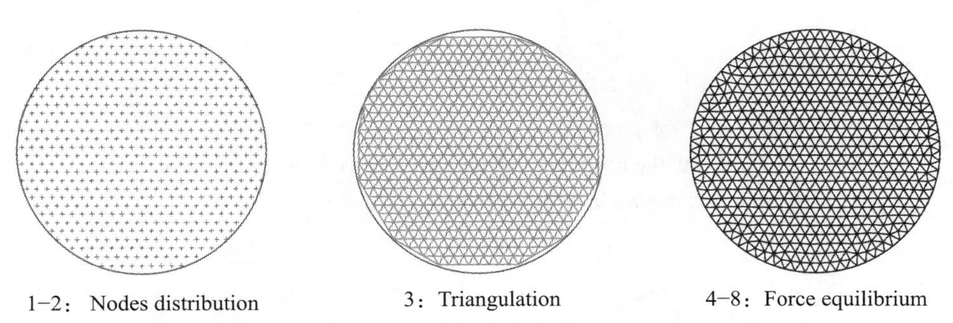

Fig. 13-2 Steps 1-8 of generating a unit circle's uniform triangular mesh

Now we describe steps 1 to 8 in the 2-D algorithm, as illustrated in Fig. 13−2 above.

Step 1. The first step creates a uniform distribution of nodes within the bounding box of the geometry, corresponding to equilateral triangles:

```
[x,y] = meshgrid(box(1,1):h0:box(2,1),box(1,2):h0*sqrt(3)/2:box(2,2));
% Fig. 13-3. (1)
x(2:2:end,:) = x(2:2:end,:) +h0/2;      % Shift even rows,see Fig. 13-3. (2)
p = [x(:),y(:)];      % List of node coordinates
```

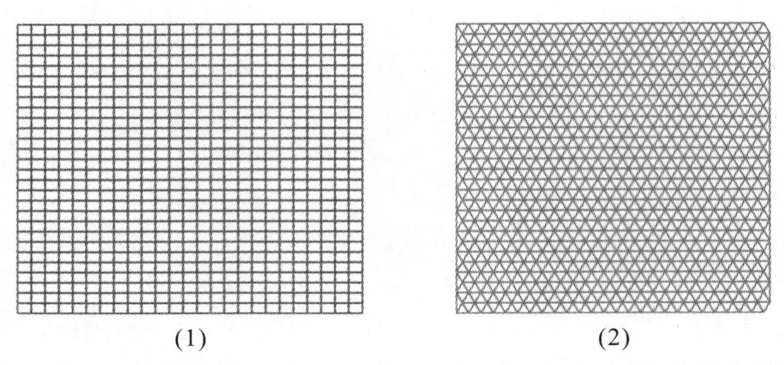

Fig. 13−3 (1) shows the initial generation of the uniform mesh in the box region roughly by using meshgrid function; (2) shifts all even rows right with the distance of $h_0/2$ to obtain a good deal of equilateral triangles

The meshgrid function generates a rectangular grid, given as two vectors x and y of node coordinates. Initially the distances are $\sqrt{3}h_0/2$ in the y-direction. By shifting every even row $h_0/2$ to the right, all points will be a distance h_0 away from their closest neighbors. The coordinates are stored in the N-by-2 array p.

Step 2. The next step is to remove all nodes outside the desired geometry (Fig. 13−4):

```
p = p(feval(fd,p,varargin{:}) < geps,:);    % Keep only d < 0 points
```

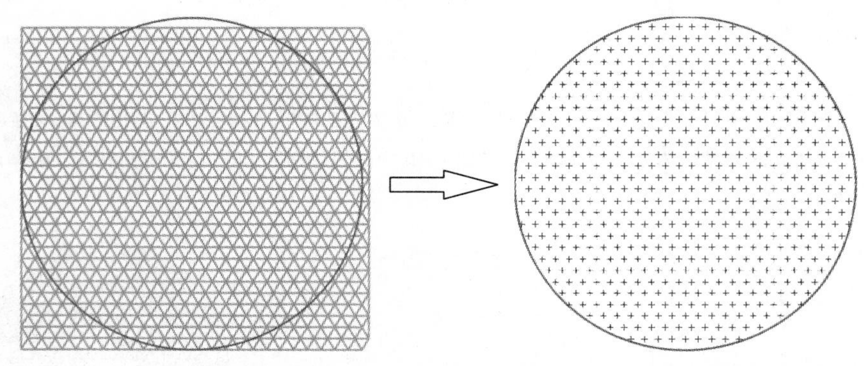

Fig. 13−4 Draw the boundaries of the structure (a unit circle in this case) we need and remove $d>0$ points which are almost out of the region

The function *feval* calls the distance function *fd*, with the node positions p and the additional arguments *varargin* as inputs. The result is a column vector of distances from the nodes to the geometry boundary. Only the interior points with negative distances (allowing a tolerance *geps*) are kept. Then we evaluate $h(x,y)$ at each node and remove points with a probability proportional to $1/[h(x,y)]^2$. In our example which is the generation of a unit circle's uniform triangular mesh, $h(x,y)$ is a constant that equals 1, so all points in the circle are kept (Fig. 13−5):

```
r0 = 1./feval(fh,p,varargin{:}).^2;       % Probability to keep point
p = [pfix; p(rand(size(p,1),1) < r0./max(r0),:)];      % Rejection method
N = size(p,1);        % Number of points N after removing
```

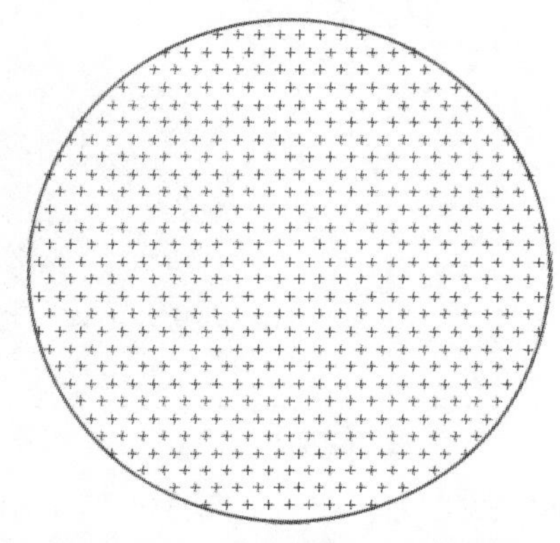

Fig. 13−5 Remove the points further according to the function $h(x, y)$ that define the probability to keep points; $h(x, y)$ represents the density variation of the mesh actually

The user's array of fixed nodes *pfix* is placed in the first few rows of *p*, *pfix* = [ ] in this case, which means all points are able to move from their original positions.

Step 3. Now the code enters the main loop, where the location of the N points is iteratively improved (Fig. 13−6). Initialize the variable *pold* for the first iteration, and start the loop (the termination criterion comes later):

```
pold = inf;
% The old points before the last iteration,the value inf is for the first iteration
while 1
...
end
```

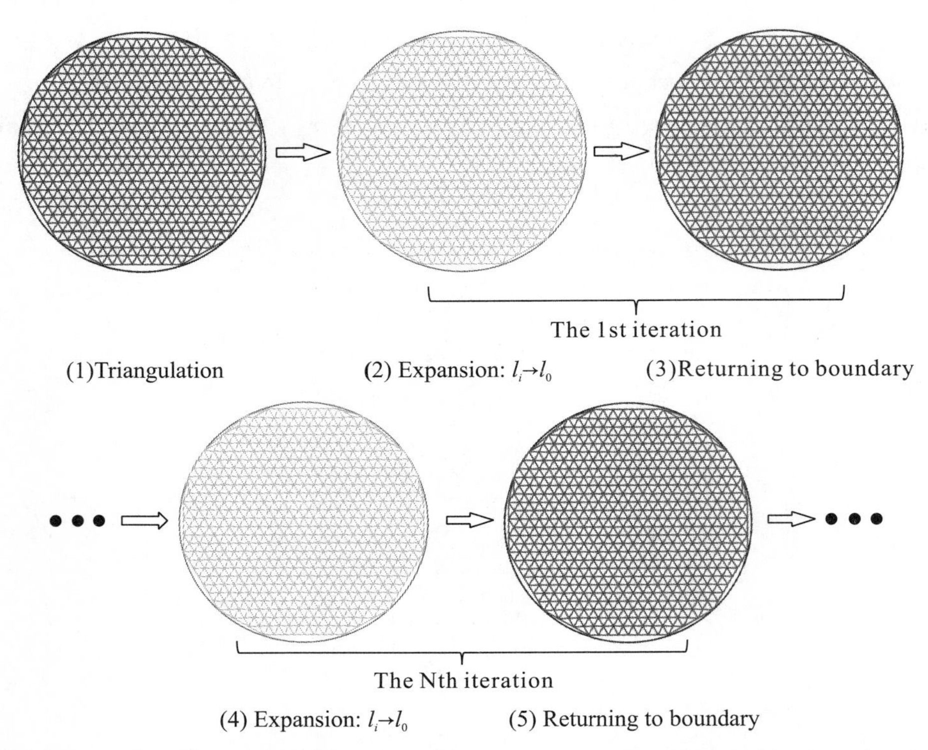

Fig. 13－6 The figure above shows the iteration process in detail. We use *delaunayn* function to triangulate these discrete points and draw the picture by taking advantage of *trimesh* function

Before evaluating the force function, a *delaunayn* triangulation should first determine the topology of the truss (Fig. 13－6 (1)). Normally this is done after the movements of nodes so that we can maintain a correct topology. To save computing time, we merely call for a retriangulation when the maximum displacement since the last triangulation is larger than *ttol* which is relative to the approximate element size l_0 :

```
if max(sqrt(sum((p - pold).^2,2))/h0) > ttol      % Any large movement?
pold = p;       % Save current positions
t = delaunayn(p);       % List of triangles
pmid = (p(t(:,1),:) + p(t(:,2),:) + p(t(:,3),:))/3;       % Compute centroids
t = t(feval(fd,pmid,varargin{:}) <- geps,:);       % Keep interior triangles
...
end
```

The node locations after retriangulation are stored in *pold*, and at the beginning of every iteration, we need to compare the relative position between the current locations *p* and *pold* to determine whether we should stop iterating. The MATLAB *delaunayn* function generates a list of triangles *t* for the set of nodes, and any triangle outside the geometry needs to be removed. We use a simple solution here—if the centroid of a triangle has $d > 0$, the triangle should be removed. This technique is not entirely robust, but it works fine in many cases, and it is very easy to implement.

Step 4. The list of triangles t is an array with 3 columns. Each row represents a triangle by three integer indices arranged in a counterclockwise order. In creating a list of edges, each triangle contributes three node pairs. Since most pairs will appear twice (two triangles share one edge), duplicates have to be removed:

```
bars = [t(:,[1,2]);t(:,[1,3]);t(:,[2,3])];      % Interior bars duplicated
bars = unique(sort(bars,2),'rows');       % Bars as node pairs
```

Step 5. The next two lines of the code give graphical output after each retriangulation. (They can be moved out of the if-statement to get more frequent output.) See the MATLAB help texts for details about these functions:

```
trimesh(t,p(:,1),p(:,2),zeros(N,1))
view(2),axis equal,axis off,drawnow
```

Step 6. Each bar is a two-component vector in *barvec*; its length is stored in L.

```
barvec = p(bars(:,1),:) - p(bars(:,2),:);      % List of bar vectors
L = sqrt(sum(barvec.^2,2));       % L = Bar lengths
```

The desired lengths L_0 come from evaluating $h(x,y)$ at the midpoint of each bar. We multiply by the scaling factor in Fig. 13−6 and the fixed factor *Fscale* to ensure that most bars give repulsive forces $f > 0$ in F.

```
hbars = feval(fh,(p(bars(:,1),:) + p(bars(:,2),:))/2,varargin{:});
% At the midpoints of the bars
L0 = hbars * Fscale * sqrt(sum(L.^2)/sum(hbars.^2));      % L0 = Desired lengths
F = max(L0 - L,0);    % Bar forces (scalars),k = 1
```

The actual update of the node positions p is in the next block of the code. The force resultant *Ftot* is the sum of force vectors in *Fvec*, from all bars meeting at a node. A stretching force has positive sign, and its direction is given by the two-component vector in bars. The sparse command is used (even though *Ftot* is immediately converted to a dense array!) because of its nice summation property for duplicated indices.

```
Fvec = F./L * [1,1].* barvec;      % Bar forces (x,y components)
```

```
Ftot = full(sparse(bars(:,[1,1,2,2]),ones(size(F)) * [1,2,1,2],
[Fvec, - Fvec],N,2));
Ftot(1:size(pfix,1),:) = 0;        % Force = 0 at fixed points
p = p + deltat * Ftot;      % Update node positions
```

Note that *Ftot* for the fixed nodes is set to zero so that their coordinates will not be changed in p .

Step 7. If a node ends up outside the geometry after the update of p, it needs to be moved back to the closest point on the boundary by using the distance function $d(x,y)$. This corresponds to a reaction force normal to the boundary. Therefore, nodes are allowed to move tangentially along the boundary. The gradient of $d(x,y)$ gives the negative direction to the closest boundary point, and it comes from numerical differentiation:

```
d = feval(fd,p,varargin{:});   ix = d > 0;        % Find points outside (d > 0)
dgradx = (feval(fd,[p(ix,1) + deps,p(ix,2)],varargin{:}) - d(ix))/deps;
dgrady = (feval(fd,[p(ix,1),p(ix,2) + deps],varargin{:}) - d(ix))/deps;
%Numerical gradient
p(ix,:) = p(ix,:) - [d(ix). * dgradx,d(ix). * dgrady];
% Project back to boundary
```

Step 8. Finally, the termination criterion is based on the maximum node movement in the current iteration (excluding the boundary nodes). We can eventually get a relatively high-quality mesh (Fig. 13-7):

```
if max(sqrt(sum(deltat * Ftot(d <- geps,:). ^2,2))/h0) < dptol,break; end
```

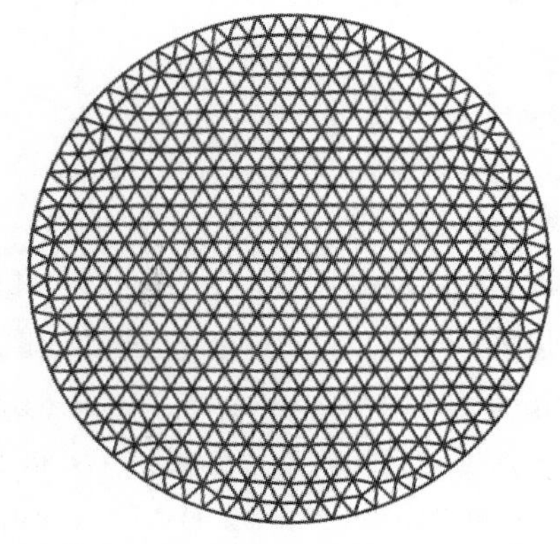

Fig. 13-7 The relatively high-quality mesh of a unit circle

This criterion is sometimes too tight, and a high-quality mesh is often achieved long before termination. In these cases, the program can be interrupted manually, or other tests can be used such as the criterion of element uniformity and so on.

13.4 Special Distance Functions

The function distmesh2d is everything that is needed to mesh a region specified by the distance $d(x,y)$ to the boundary. While it is easy to create distance functions for some simple geometries, we still need to define some short help functions for more complex geometries.

For example, the output of the function *dcircle* is the signed distance from p to a circle with center (xc, yc) and radius r. As for a rectangle, we use the function *drectangle* to express the minimum signed distance from the nodes to the four boundary lines of the rectangle. Of course, if a node is in the rectangle, the distance sign is negative. What we should note is that it is not the completely correct distance to the four external regions whose nearest points are corners of the rectangle. The function avoids square roots from distances to corner points, and no mesh points end up in these four regions when the corner points are fixed (by $pfix$). However, this makes our algorithm easier and more efficient.

The functions *dunion*, *ddiff* and *dintersect* can combine two geometries like Boolean operations. Here we use two separate projections to the regions A and B corresponding to the distances $dA(x,y)$ and $dB(x,y)$:

$$Union: dA \cup B(x,y) = \min(dA(x,y), dB(x,y)) \tag{13-7}$$

$$Difference: dA \setminus B(x,y) = \max(dA(x,y), -dB(x,y)) \tag{13-8}$$

$$Intersection: dA \cap B(x,y) = \max(dA(x,y), dB(x,y)) \tag{13-9}$$

Finally, the functions *pshift* and *protate* can operate on the node array p to translate or rotate the coordinates.

Short help functions for the generation of distance functions and size functions are shown as below:

```
function  d = dcircle(p,xc,yc,r)     % Circle
d = sqrt((p(:,1) - xc).^2 + (p(:,2) - yc).^2) - r;

function  d = drectangle(p,x1,x2,y1,y2)     % Rectangle
d = -min(min(min(-y1 + p(:,2),y2 - p(:,2)), -x1 + p(:,1)),x2 - p(:,1));

function  d = dunion(d1,d2)     % Union
d = min(d1,d2);
```

```
function d = ddiff(d1,d2)             % Difference
d = max(d1, -d2);

function d = dintersect(d1,d2)        % Intersection
d = max(d1,d2);

function p = pshift(p,x0,y0)          % Shift points
p(:,1) = p(:,1) - x0;
p(:,2) = p(:,2) - y0;

function p = protate(p,phi)           % Rotate points around origin
A = [cos(phi), - sin(phi); sin(phi), cos(phi)]; p = p * A;

function h = huniform(p,varargin)     % Uniform h(x,y) distribution
h = ones(size(p,1),1);
```

13.5 Examples

Fig. 13-8 shows a number of examples, starting from a circle and extending to relatively complicated meshes.

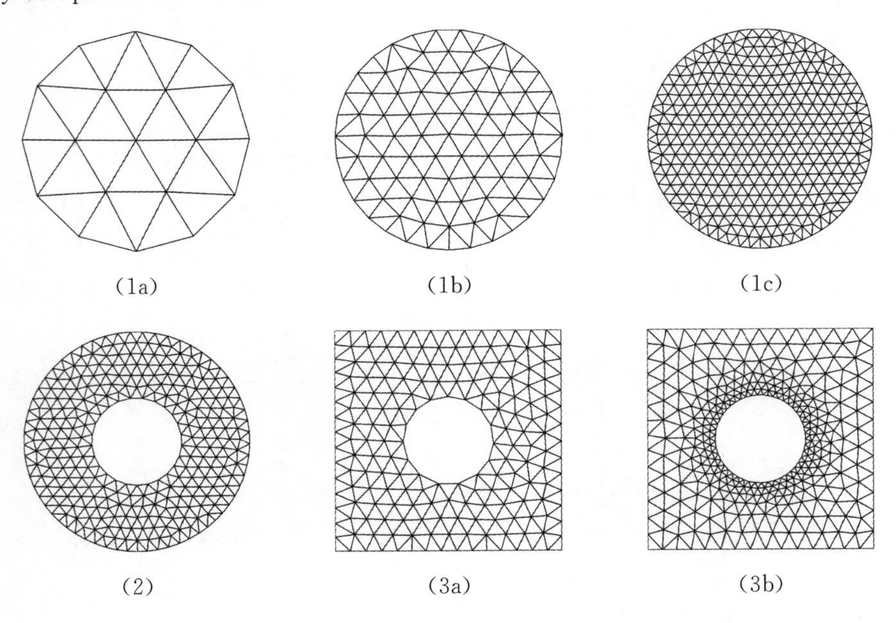

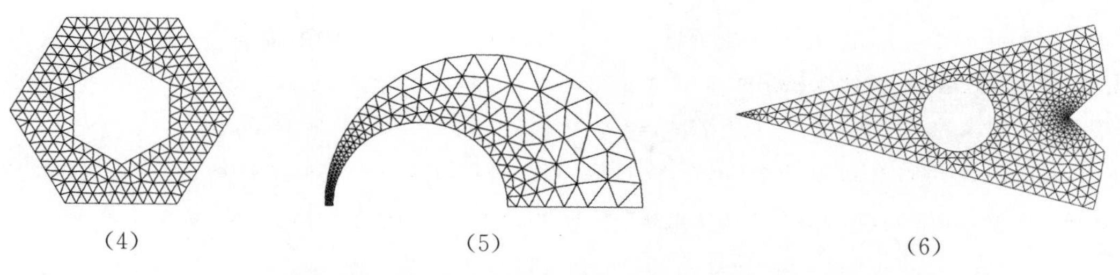

(4) (5) (6)

Fig. 13−8 Example meshes are shown above; Particularly, examples (3b), (5) and (6) have varying size functions $h(x, y)$

13.5.1 Unit Circle

We work directly with $d = \sqrt{x^2 + y^2} - 1$, which can be specified as an inline function. For a uniform mesh, $h(x, y)$ returns a vector of 1's. The circle has bounding box $-1 \leqslant x \leqslant 1$, $-1 \leqslant y \leqslant 1$, with no fixed points. A mesh with element size approximately $h_0 = 0.2$ is generated with two lines of the code:

```
fd = inline('sqrt(sum(p.^2,2)) - 1','p');
[p,t] = distmesh2d(fd,@huniform,0.2,[-1,-1;1,1],[]);
```

The plots (1a), (1b) and (1c) show the resulting meshes for $h_0 = 0.4$, $h_0 = 0.2$ and $h_0 = 0.1$. Inline functions are defined without creating a separate file. The first argument is the function itself, and the remaining arguments name the parameters to the function (help *inline* brings more information). Please note the comment near the end of the paper about the relatively slow performance of inline functions.

13.5.2 Unit Circle with Hole

Removing a circle of radius 0.4 from the unit circle gives the distance function $d(x, y) = \left|0.7 - \sqrt{x^2 + y^2}\right| - 0.3$. The function means that the distance from those points to the midline of the ring is no greater than 0.3.

```
fd = inline('-0.3 + abs(0.7 - sqrt(sum(p.^2,2)))');
[p,t] = distmesh2d(fd,@huniform,0.1,[-1,-1;1,1],[]);
```

Equivalently, $d(x, y)$ is the distance to the difference of two circles:

```
fd = inline('ddiff(dcircle(p,0,0,1),dcircle(p,0,0,0.4))','p');
```

13.5.3 Square with Hole

We can replace the outer circle with a square, keeping the circular hole. Since our distance function *drectangle* is incorrect at the corners, so we fix those four nodes:

```
fd = inline('ddiff(drectangle(p, -1,1, -1,1),dcircle(p,0,0,0.4))','p');
pfix = [-1, -1; -1,1;1, -1;1,1];
[p,t] = distmesh2d(fd,@huniform,0.15,[-1, -1;1,1],pfix);
```

A non-uniform $h(x,y)$ gives a finer resolution close to the circle (Fig. 13−8(3b)):

```
fh = inline('min(4*sqrt(sum(p.^2,2)) -1,2)','p');
[p,t] = distmesh2d(fd,fh,0.05,[-1, -1;1,1],pfix);
```

13.5.4 Polygons

It is easy to create *dpoly* for the distance to a given polygon, using the MATLAB function *inpolygon* to determine the sign. We mesh a regular hexagon and fix its six corners:

```
phi = (0:6)'/6*2*pi;
pfix = [cos(phi),sin(phi)];
[p,t] = distmesh2d(@dpoly,@huniform,0.1,[-1, -1;1,1],pfix,pfix);
```

Note that *pfix* is passed twice, the first one is to specify the fixed points, and the second one is a parameter of *dpoly* to specify the polygon. In Fig. 13−8(4), we also removed a smaller rotated hexagon by using *ddiff*.

13.5.5 Geometric Adaptivity

Here we show how the distance functions work in the definition of $h(x,y)$ to implement the expression of local feature size by using adaptive logic. The half-plan $y>0$ has $d(x,y)=-y$, and our $d(x,y)$ is created by an intersection and a difference:

$$d_1=\sqrt{x^2+y^2}-1 \tag{13-10}$$

$$d_2=\sqrt{(x+0.4)^2+y^2}-0.55 \tag{13-11}$$

$$d=\max(d_1,-d_2,-y) \tag{13-12}$$

Next, we create two element size functions $h_1(x,y)$ and $h_2(x,y)$ to represent the finer resolutions near the circles. The element sizes $h_1(x,y)$ and $h_2(x,y)$ increase with the distances from the boundaries:

$$h_1(x,y) = 0.15 - 0.2d_1(x,y) \tag{13-13}$$

$$h_2(x,y) = 0.06 + 0.2d_2(x,y) \tag{13-14}$$

These are made proportional to the two radii to get equal angular resolutions. Note the minus sign for d_1 since it is negative inside the region. As for the local feature size in the middle of boundaries, we resolve it by defining the third size function $h_3(x,y)$:

$$h_3(x,y) = \frac{d_2(x,y) - d_1(x,y)}{3} \tag{13-15}$$

Finally, the three size functions are combined to yield the mesh in Fig. 13-8 (5):

$$h = \min(h_1, h_2, h_3) \tag{13-16}$$

The initial distribution had size $h_0 = 0.05/3$ and four fixed corner points.

13.5.6 More complex geometry

The example (6) shows a somewhat more complicated construction, involving set operations on circles and rectangles, and element sizes increasing away from two vertices and the circular hole.

13.6 Mesh Generation in 3-D

Many scientific and engineering simulations require 3-D modeling. The boundaries become surfaces, and the interior becomes a volume instead of an area. For example, the convex hull of a 3-D model is a convex polytope rather than a polygon, and it is tessellated by tetrahedra rather than triangles. In a 3-D model, the mesh generator function is *distmeshnd*. The truss lies in a higher-dimensional space, and each simplex has 6 edges (compared to three for triangles). The distribution uses a regular grid. The input node array *p* to *delaunay* is *N*-by-3. The post-processing of a tetrahedral mesh is somewhat different, but the MATLAB visualization routines make this relatively easy as well.

In 2-D we usually fix all the corner points, when the distance functions are not accurate close to corners. In 3-D, we would have to fix points along intersections of surfaces. A choice of edge length along those curves might be difficult for nonuniform meshes. An alternative is to generate 'correct' distance functions, without the simplified assumptions in *drectangle*, *dunion*, *ddiff*, and *dintersect*. It is able to handle all convex intersections, and the technique is used in the cylinder example below.

Now, we generate the tetrahedral meshes in Fig. 13-9.

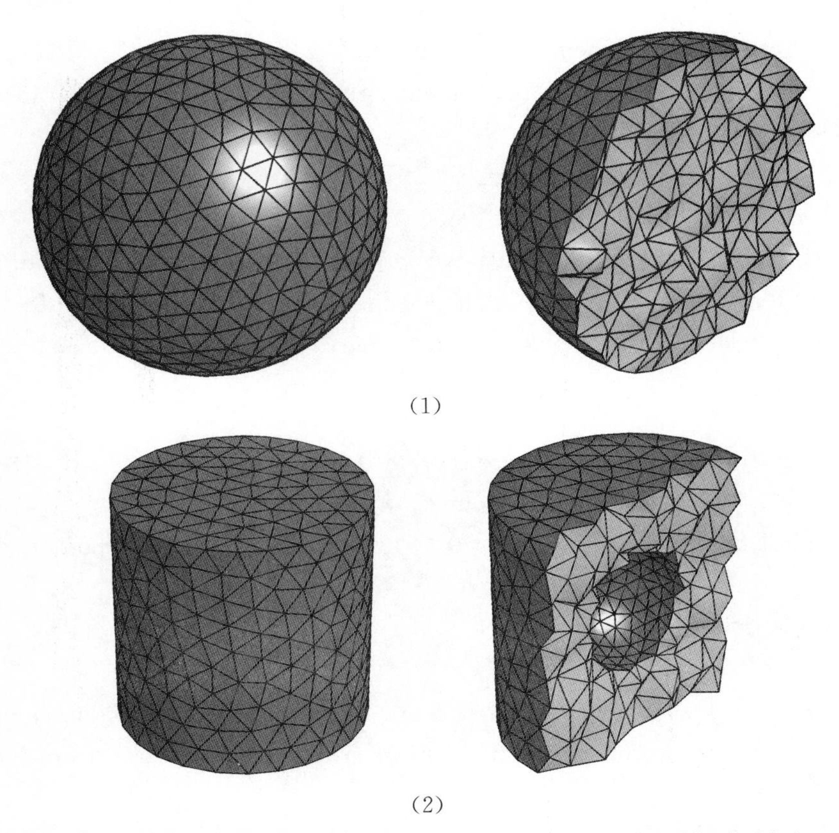

(1)

(2)

Fig. 13—9 Tetrahedral meshes of a ball and a cylinder with a spherical hole; The left plots show the surface meshes, and the right plots show cross-sections

13.6.1 Unit Ball

The ball in 3-D uses nearly the same code as the circle:

```
fd = inline('sqrt(sum(p.^2,2)) - 1','p');
[p,t] = distmeshnd(fd,@huniform,0.15,[-1, -1, -1;1,1,1],[]);
```

This distance function fd automatically sums over three dimensions, and the bounding box has two more components. The resulting mesh has 1295 nodes and 6349 tetrahedra.

13.6.2 Cylinder with Spherical Hole

Fora cylinder with radius 1 and height 2, we create d_1, d_2, d_3, for the curved surface and the top and bottom:

$$d_1(x,y,z) = \sqrt{x^2 + y^2} - 1 \tag{13-17}$$

$$d_2(x,y,z) = z - 1 \tag{13-18}$$

$$d_3(x,y,z) = -z - 1 \tag{13-19}$$

An approximate distance function is then formed by intersection:

$$d_{\approx} = \max(d_1 + d_2 + d_3) \tag{13-20}$$

This would be sufficient if the 'corner points' along the curves $x_2 + y_2 = 1$, $z = \pm 1$ were fixed by an initial node placement. Better results can be achieved by correcting the distance function using distances to the two curves:

$$d_4(x,y,z) = \sqrt{[d_1(x,y,z)]^2 + [d_2(x,y,z)]^2} \tag{13-21}$$

$$d_5(x,y,z) = \sqrt{[d_1(x,y,z)]^2 + [d_3(x,y,z)]^2} \tag{13-22}$$

These functions should be used where the intersections of d_1, d_2, and d_1, d_3 overlap, that is, when they both are positive:

$$d = \begin{cases} d_4, & \text{if } d_1 > 0 \text{ and } d_2 > 0 \\ d_5, & \text{if } d_1 > 0 \text{ and } d_3 > 0 \\ d_{\approx}, & \text{otherwise} \end{cases} \tag{13-23}$$

Fig. 13−9 (2) shows a mesh for the difference between this cylinder and a ball of radius 0. 5. We define $h(x,y,z) = \min(4\sqrt{x^2 + y^2 + z^2} - 1, 2)$ and $h_0 = 0.1$ to obtain a finer resolution close to this ball, and the last mesh has 1057 nodes and 4539 tetrahedra.

Chapter 14 Model Generation in ANSYS

14. 1 Understanding Model Generation

Structural analysis is probably the most common application of the finite element method. The term structure implies not only civil engineering structures such as bridges and buildings, but also naval, aeronautical, and mechanical structures such as ship hulls, aircraft bodies, and machine housings, as well as mechanical components such as pistons, machine parts, and tools. The ultimate purpose of a finite element analysis is to recreate mathematically the behavior of an actual engineering system. In other words, the analysis must be an accurate mathematical model of a physical prototype. In the broadest sense, this model comprises all the nodes, elements, material properties, real constants, boundary conditions, and other features that are used to represent the physical system.

14. 1. 1 What is Model Generation

In ANSYS terminology, model generation usually takes on the narrower meaning of generating the nodes and elements that represent the spatial volume and connectivity of the actual system. Thus, model generation in this discussion means the process of defining the geometric configuration of the model's nodes and elements. The ANSYS program offers the following approaches to model generation:

- Creating a solid model within ANSYS.
- Using direct generation.
- Importing a model created in a computer-aided design (CAD) system.

14. 1. 2 Typical Steps Involved in Model Generation Within ANSYS

A common modeling session might follow this general outline (detailed information on italicized subjects can be found elsewhere in this guide):

- Begin by planning your approach. Determine your objectives, decide what basic form your model will take, choose appropriate element types, and consider how you will establish an appropriate mesh density. You will typically do this general planning before you initiate your ANSYS session.

- Enter the preprocessor (**PREP7**) to initiate your model-building session. Most often, you will build your model using solid modeling procedures.
- Establish a working plane.
- Generate basic geometric features using geometric primitives and Boolean operators.
- Activate the appropriate coordinate system.
- Generate other solid model features from the bottom up. That is, create key points, and then define lines, areas, and volumes as needed.
- Use more Boolean operators or number controls to join separate solid model regions together as appropriate.
- Create tables of element attributes (element types, real constants, material properties, and element coordinate systems).
- Set element attribute pointers.
- Set meshing controls to establish your desired mesh density if desired. This step is not always required because default element sizes exist when you enter the program. (If you want the program to refine the mesh automatically, exit the preprocessor at this point, and activate adaptive meshing.)
- Create nodes and elements by meshing your solid model.
- After you have generated nodes and elements, add features such as surface-to-surface contact elements, coupled degrees of freedom, and constraint equations.
- Save your model data to Jobname. db.
- Exit the preprocessor.

The solid modeling features of ANSYS are known to have robustness issues. By careful planning and use of alternative strategies, you can successfully create the model required for analysis. However, you may be better served using your CAD modeler to create your model or using Design Modeler under the ANSYS Workbench environment to create your model.

14. 2 Planning Your Approach

As you begin to create your model, you will make a number of decisions that determine how you will mathematically simulate the physical system. For example: What are the objectives of your analysis? Will you model all, or just a portion, of the physical system? How much detail will you include in your model? What kinds of elements will you use? How dense should your finite element mesh be?

In general, you will attempt to balance computational expense (such as CPU time) against precision of results as you answer these questions. The decisions you make in the planning stage of your analysis will largely govern the success or failure of your analysis efforts.

This first step of your analysis relies not on the capabilities in the ANSYS program, but on your own education, experience, and professional judgment. Only you can determine what the objectives of your analysis must be. The objectives you establish at the start will influence the remainder of your choices as you generate the model.

14.3 Choosing a Model Type (2-D, 3-D, etc.)

Your finite element model may be categorized as being 2-D or 3-D, and as being composed of point elements, line elements, area elements, or solid elements. Of course, you can intermix different kinds of elements as required (taking care to maintain the appropriate compatibility among degrees of freedom). For example, you might model a stiffened shell structure using 3-D shell elements to represent the skin and 3-D beam elements to represent the ribs. Your choice of model dimensionality and element type will often determine which method of model generation will be most practical for your problem.

LINE models can represent 2-D or 3-D beam or pipe structures, as well as 2-D models of 3-D axisymmetric shell structures. Solid modeling usually does not offer much benefit for generating line models; they are more often created by direct generation methods. 2-D SOLID analysis models are used for thin planar structures (plane stress), 'infinitely long' structures having a constant cross section (plane strain), or axisymmetric solid structures. Although many 2-D analysis models are relatively easy to create by direct generation methods, they are usually easier to create with solid modeling.

3-D SHELL models are used for thin structures in 3-D space. Although some 3-D shell analysis models are relatively easy to create by direct generation methods, they are usually easier to create with solid modeling.

3-D SOLID analysis models are used for thick structures in 3-D space that have neither a constant cross section nor an axis of symmetry. Creating a 3-D solid analysis model by direct generation methods usually requires considerable effort. Solid modeling will nearly always make the job easier.

14.4 Choosing Between Linear and Higher Order Elements

The ANSYS program's element library includes two basic types of area and volume elements: linear (with or without extra shapes), and quadratic. These basic element types are represented schematically in Fig. 14－1. Let's examine some of the considerations involved in choosing between these two basic element types:

(a) Linear isoparametric;

(b) Linear isoparametric with extra shapes;

(c) Quadratic.

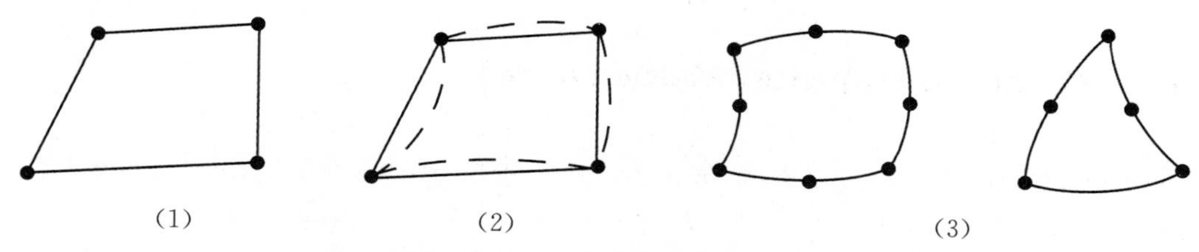

Fig. 14－1 Area and Volume Types

14.4.1 Linear Elements (No Midside Nodes)

For structural analyses, these corner node elements with extra shape functions will often yield an accurate solution in a reasonable amount of computer time. When using these elements, it is important to avoid their degenerate forms in critical regions. That is, avoid using the triangular form of 2-D linear elements and the wedge or tetrahedral forms of 3-D linear elements in high results-gradient regions, or other regions of special interest. You should also take care to avoid using excessively distorted linear elements. In nonlinear structural analyses, you will usually obtain better accuracy at less expense if you use a fine mesh of these linear elements rather than a comparable coarse mesh of quadratic elements. Examples of (1) linear and (2) quadratic elements are shown in Fig. 14－2.

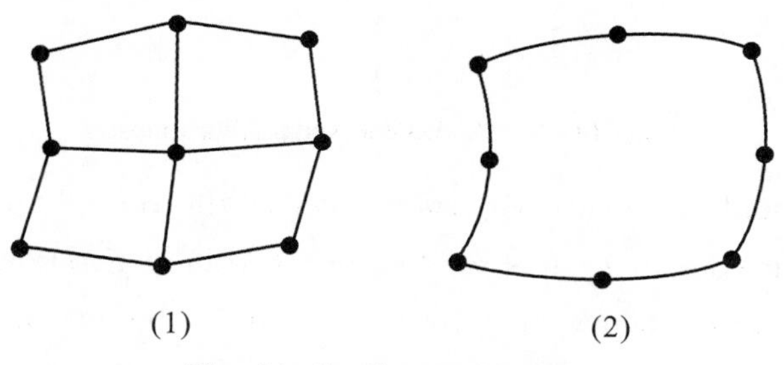

Fig. 14－2 Comparable Grids

When modeling a curved shell, you must choose between using curved (that is, quadratic) or flat (linear) shell elements. Each choice has its advantages and disadvantages. For most practical cases, the majority of problems can be solved to a high degree of accuracy in a minimum amount of computer time with flat elements. You must take care, however, to ensure that you use enough flat elements to model the curved surface adequately. Obviously, the smaller the element, the better the accuracy. It is recommended that the 3-D flat shell elements not extend over more than a 15° arc. Conical shell (axisymmetric line) elements should be limited to a 10° arc (or 5° if near the y axis).

For most non-structural analyses (thermal, magnetic, etc.), the linear elements are nearly as good as the higher order elements, and are less expensive to use. Degenerate elements (triangles and tetrahedra) usually produce accurate results in non-structural analyses.

14.4.2 Quadratic Elements (Midside Nodes)

For linear structural analyses with degenerate element shapes (that is, triangular 2-D elements and wedge or tetrahedral 3-D elements), the quadratic elements will usually yield better results at less expense than will the linear elements. However, in order to use these elements correctly, you need to be aware of a few peculiar traits that they exhibit:

- Distributed loads and edge pressures are not allocated to the element nodes according to 'common sense', as they are in the linear elements. (See Fig. 14－3) Reaction forces from midside-node elements exhibit the same nonintuitive interpretation.

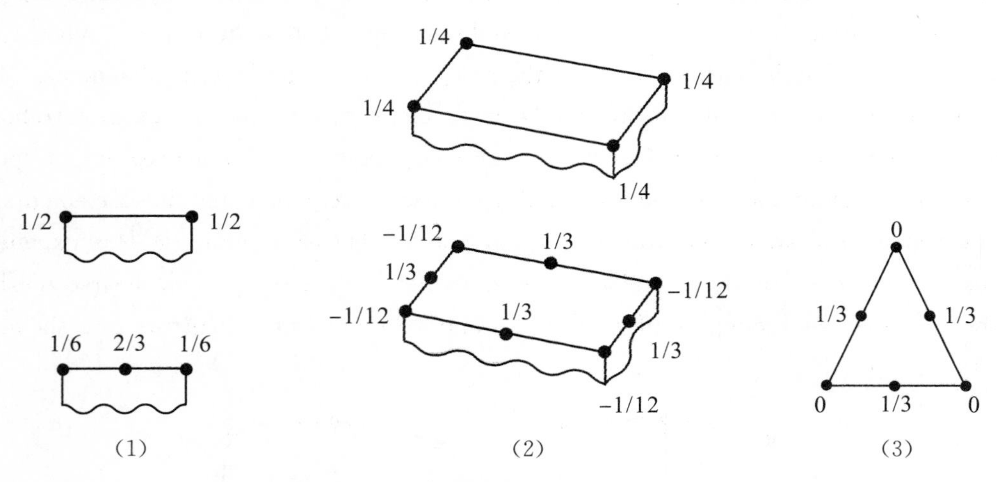

Fig. 14－3 Equivalent Nodal Allocations

- Mass at the midside nodes is greater than at the corner nodes. When selecting master degrees of freedom in a substructure (or CMS) generation, you must include midside nodes as master nodes in order to achieve better mass or surface load representation.

• In dynamic analyses where wave propagation is of interest, midside-node elements are not recommended because of the nonuniform mass distribution.

• When constraining degrees of freedom at an element edge (or face), all nodes on the face, including the midside nodes, must be constrained.

• The corner node of an element should only be connected to the corner node, and not the midside node of an adjacent element. Adjacent elements should have connected (or common) midside nodes.

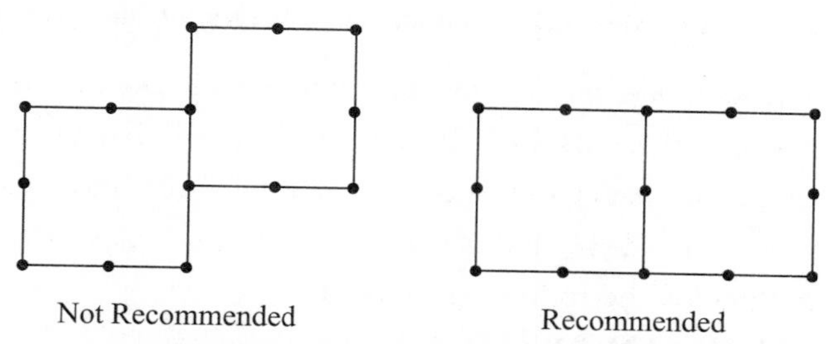

Fig. 14−4 Avoid Midside-to-Corner Node Connections Between Elements

• For elements having midside nodes, it is generally preferred that each such node be located at the straight-line position halfway between the corresponding corner nodes. There are, however, situations where other locations may be more desirable:

— Nodes following curved geometric boundaries will usually produce more accurate analysis results and all ANSYS meshers place them there by default.

— Even internal edges in some meshes may have to curve to prevent elements from becoming inverted or otherwise overly distorted. ANSYS meshers sometimes produce this type of curvature.

— It is possible to mimic a crack-tip singularity with 'quarter point' elements, with midside nodes deliberately placed off-center. You can produce this type of specialized area mesh in ANSYS by using the **KSCON** command (**Main Menu>Preprocessor>Meshing>Size Cntrls>Concentrat KPs>Create**).

• If you do not assign a location for a midside node, the program will automatically place that node midway between the two corner nodes, based on a linear Cartesian interpolation. Nodes located in this manner will also have their nodal coordinate system rotation angles linearly interpolated.

• Connecting elements should have the same number of nodes along the common side. When mixing element types it may be necessary to remove the midside node from an element. For example, node N of the 8-node element shown below should be removed (or given a zero-node number when the element is created) when the element is connected to a 4-node element.

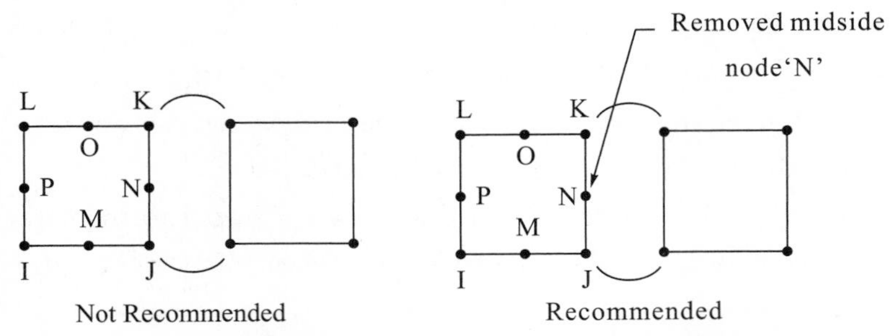

Fig. 14—5 Avoid Mismatched Midside Nodes at Element Interconnections

• A removed midside node implies that the edge is and remains straight, resulting in a corresponding increase in the stiffness. It is recommended that elements with removed nodes be used only in transition regions and not where simpler linear elements with added shape functions will do. If needed, nodes may be added or removed after an element has been generated, using one of the following methods:

Command (s): **EMID, EMODIF**

— **GUI: Main Menu>Preprocessor>Modeling>Move/Modify>Elements>Add Mid Nodes**

— **Main Menu>Preprocessor>Modeling>Move/Modify>Elements>Remove Mid Nd**

— **Main Menu>Preprocessor>Modeling>Move/Modify>Elements>Modify Nodes**

• A quadratic element has no more integration points than a linear element. For this reason, linear elements will usually be preferred for nonlinear analyses.

• In postprocessing, the program uses only corner nodes for section and hidden line displays. Similarly, nodal stress data for printout and postprocessing are available only for the corner nodes.

14.5 Solid Modeling and Direct Generation

You can use two different methods to generate your model: solid modeling and direct generation. With solid modeling, you describe the geometric boundaries of your model, establish controls over the size and desired shape of your elements, and then instruct the ANSYS program to generate all the nodes and elements automatically. By contrast, with the direct generation method, you determine the location of every node and the size, shape, and connectivity of every element prior to defining these entities in your ANSYS model.

Although some automatic data generation is possible, the direct generation method is essentially a hands-on, 'manual' method that requires you to keep track of all your node numbers as you develop your finite element mesh. This detailed bookkeeping can become tedious for large models, contributing to the potential for modeling errors. Solid modeling is usually more powerful and versatile than direct generation, and is commonly the preferred method for generating your model.

In spite of the many advantages of solid modeling, you might occasionally encounter circumstances where direct generation will be more useful. You can easily switch back and forth between direct generation and solid modeling, using the different techniques as appropriate to define different parts of your model. Detailed discussions of solid modeling and direct generation can be found in Solid Modeling and Direct Generation, respectively. To help you judge which method might be more suitable for a given situation, the relative advantages and disadvantages of the two approaches are summarized here.

14.5.1 Solid Modeling

The purpose of using a solid model is to relieve you of the time-consuming task of building a complicated finite element model by direct generation. Some solid modeling and meshing operations can help you to speed up the creation of your final analysis model.

The solid modeling features of ANSYS are known to have robustness issues. With careful planning and alternative strategies, you can successfully create the model required for analysis. However, you may be better served using your CAD modeler or ANSYS Design Modeler under the ANSYS Workbench environment to create your model.

14.5.1.1 Creating Your Solid Model from the Bottom Up

The points that define the vertices of your model are called keypoints and are the 'lowest-order' solid model entities. If in building your solid model, you first create your keypoints, and then use those keypoints to define the 'higher-order' solid model entities (that is, lines, areas, and volumes), you are said to be building your model 'from the bottom up'. Models built from the bottom up are defined within the currently active coordinate system.

Keypoints are the vertices, lines are the edges, areas are the faces, and volumes are the interior of the object. Notice that there is a hierarchy in these entities: volumes, the highest-order entities, are bounded by areas, which are bounded by lines, which in turn are bounded by keypoints. In bottom up construction, you first create keypoints and use those keypoints to define higher-order solid model entities (lines, areas and volumes).

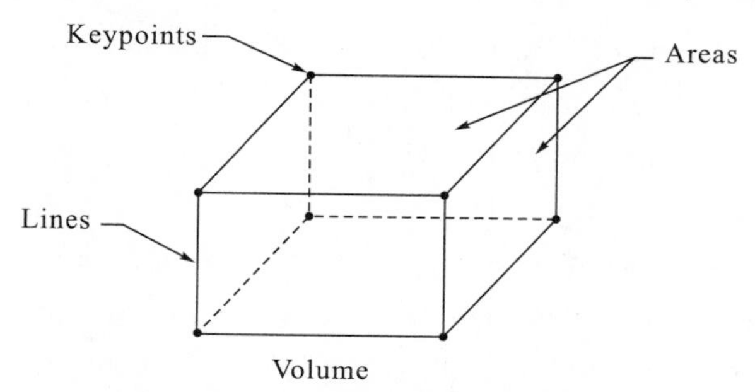

Fig. 14—6 Bottom Up Construction

14.5.1.2 Creating Your Solid Model from the Top Down: Primitives

In top down construction, you use geometric primitives (fully-defined lines, areas and volumes) to assemble your model. As you create a primitive, the program automatically creates all the 'lower' entities associated with it. A geometric primitive is a commonly used solid modeling shape (such as a sphere or regular prism) that can be created with a single ANSYS command.

Because primitives are higher-order entities that can be constructed without first defining any keypoints, model generation that uses primitives is sometimes referred to as 'top down' modeling. (When you create a primitive, the program automatically creates all the necessary lower-order entities, including keypoints.) Geometric primitives are created within the working plane. You can freely combine bottom up and top down modeling techniques, as appropriate, in any model. Remember that geometric primitives are built within the working plane while bottom up techniques are defined against the active coordinate system. If you are mixing techniques, you may wish to consider using the CSYS, WP or CSYS, 4 command to force the coordinate system to follow the working plane.

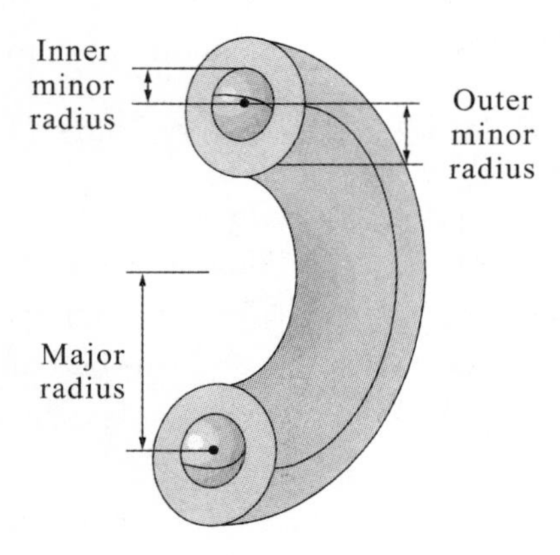

Fig. 14—7 Top Down Constructions (Primitives)

14.5.1.3 Sculpting Your Model with Boolean Operations

Boolean algebra provides a means for combining sets of data, using such logical operators as intersect, union, subtract, etc. The ANSYS program allows you to apply these same Boolean operators to your solid model, so that you can modify your solid model constructions more easily.

You can apply Boolean operations to almost any solid model construction, whether it was created from the top down or from the bottom up. The only exceptions are that Boolean operations are not valid for entities created by concatenation and that some Boolean operations cannot always be performed on entities that contain degeneracies.

Also, all solid-model loads and element attributes should be defined after you complete your Boolean operations. If you are using Booleans to modify an existing model, you should take care to redefine your element attributes and solid-model loads.

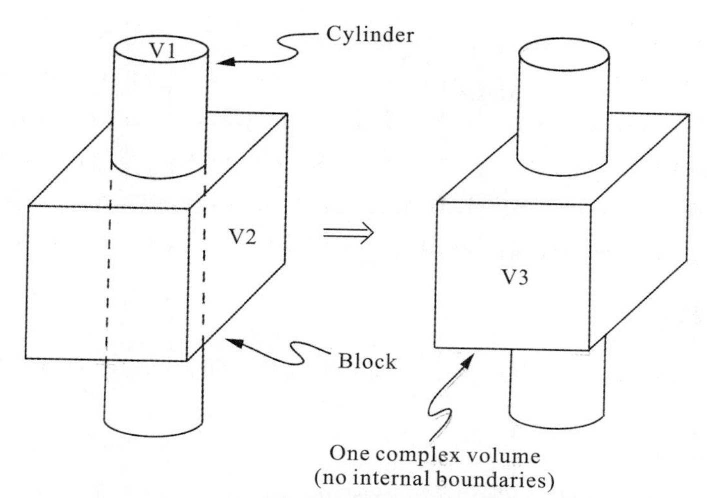

Fig. 14−8 Create Complex Shapes With Boolean Operations

14. 5. 1. 4 Moving and Copying Solid Model Entities

If your model repetitively uses a relatively complicated area or volume, you need construct that part only once; you can then generate copies of that part in new locations and new orientations as needed. For example, the elongated voids in the plate shown below can be copied from a single such void.

Geometric primitives can also be considered to be 'parts'. As you create geometric primitives, their location and orientation will be determined by the current working plane. Because it is not always particularly convenient to redefine the working plane for each new primitive that you create, you might find it more practical to allow a primitive to be created at the 'wrong' location, and then move that primitive to its correct position. Of course, this operation is not limited to geometric primitives: any solid model entity can be copied or moved.

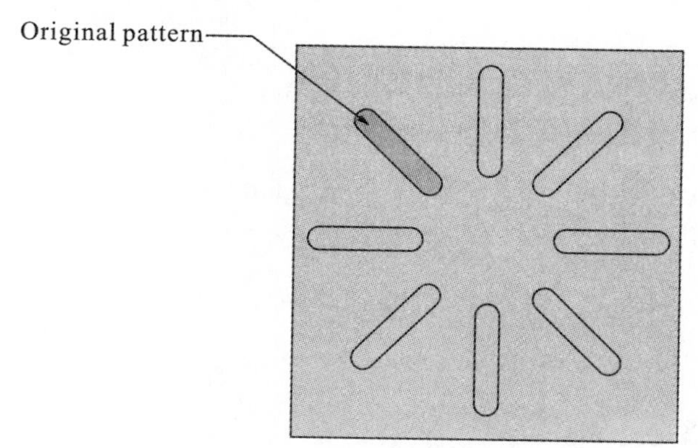

Fig. 14−9 Copying an Area

14. 5. 2 Direct Generation

Direct generation is the approach in which you define the nodes and elements of a model directly. Despite the many convenience commands that allow you to copy, reflect, scale, etc. a given pattern of nodes or elements, direct generation can commonly require about ten times as many data entries to define a model as compared to solid modeling.

A model assembled by direct generation is defined strictly in terms of nodes and elements. Even though node and element generation operations can be interspersed, no one element can be defined until after all of its nodes have been created.

14. 5. 2. 1 Nodes

This section describes various tasks related to the direct generation of nodes. Topics include:

- Defining nodes.
- Generating additional nodes from existing nodes.
- Maintaining (list, view and delete) nodes.
- Moving nodes.
- Rotating a node's coordinate system.
- Reading and writing text files that contain nodal data.

14. 5. 2. 2 Elements

This section describes various tasks related to the direct generation of elements. Topics include:

- Prerequisites for defining elements.
- Assembling element tables.
- Pointing to entries in element tables.
- Reviewing the contents of element tables.
- Defining elements.
- Maintaining (list, view, and delete) elements.
- Generating additional elements from existing elements.
- Using special methods for generating elements.
- Reading and writing text files that contain element data.
- Modifying elements by changing nodes.
- Modifying elements by changing element attributes.

14. 6 Generating the Mesh

The process for generating a mesh of nodes and elements consists of three general steps:

1. Set the element attributes.

2. Set mesh controls (optional). A large number of mesh controls are available from which to choose.

3. Meshing the model.

It is not always necessary to set mesh controls because the default mesh controls are appropriate for many models. If no controls are specified, the program will use the default settings (**DESIZE**) to produce a free mesh. Alternatively, you can use the SmartSize feature to produce a better quality free mesh.

14.6.1 Free or Mapped Mesh

Before meshing the model, and even before building the model, it is important to think about whether a free mesh or a mapped mesh is appropriate for the analysis. A free mesh has no restrictions in terms of element shapes, and has no specified pattern applied to it. A mapped mesh is restricted in terms of the element shape it contains and the pattern of the mesh. A mapped area mesh contains either only quadrilateral or only triangular elements, while a mapped volume mesh contains only hexahedron elements. In addition, a mapped mesh typically has a regular pattern, with obvious rows of elements. If you want this type of mesh, you must build the geometry as a series of fairly regular volumes and/or areas that can accept a mapped mesh.

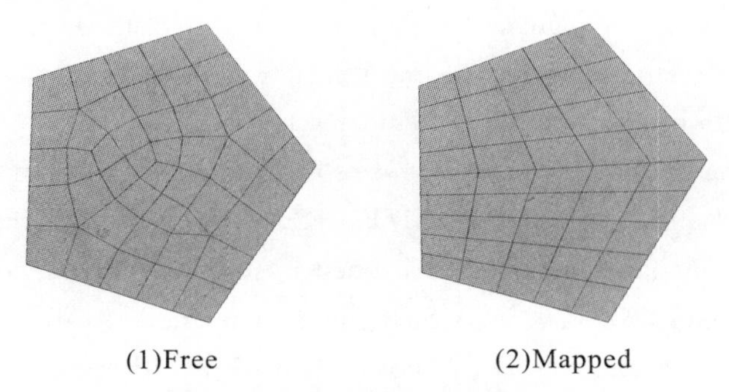

(1)Free (2)Mapped

Fig. 14—10 Free and Mapped Meshes

You use the MSHKEY command or the equivalent GUI path (both of which are described later) to choose a free or a mapped mesh. Keep in mind that the mesh controls you use will vary depending on whether a free or mapped mesh is desired.

14.6.2 Setting Element Attributes

Before you generate a mesh of nodes and elements, you must first define the appropriate element attributes. That is, you must specify the following:

- The element type.
- Real constant set (usually comprising the element's geometric properties, such as thickness or cross-sectional area).

- Material properties set (such as Young's modulus, thermal conductivity, etc.).
- Element coordinate system.
- Section ID.

14.6.3 Mesh Controls

The default mesh controls that the program uses may produce a mesh that is adequate for the model you are analyzing. In this case, you will not need to specify any mesh controls. However, if you do use mesh controls, you must set them before meshing your solid model.

Mesh controls allow you to establish such factors as the element shape, midside node placement, and element size to be used in meshing the solid model. This step is one of the most important of your entire analysis, for the decisions you make at this stage in your model development will profoundly affect the accuracy and economy of your analysis.

14.6.3.1 Controls Used for Free Meshing

In free meshing operations, no special requirements restrict the solid model. Any model geometry, even if it is irregular, can be meshed. The element shapes used will depend on whether you are meshing areas or volumes. For area meshing, a free mesh can consist of only quadrilateral elements, only triangular elements, or a mixture of the two. For volume meshing, a free mesh is usually restricted to tetrahedral elements. Pyramid-shaped elements may also be introduced into the tetrahedral mesh for transitioning purposes.

For free meshing operations, element sizes are produced based on the current settings of the **DESIZE** command, along with **ESIZE**, **KESIZE**, and **LESIZE.** If SmartSizing is turned on, the element sizes will be determined by the **SMRTSIZE** command along with **ESIZE**, **KESIZE**, and **LESIZE.** (SmartSizing is recommended for free meshing.) You can find all of these meshing controls under both **Main Menu>Preprocessor>Meshing>Mesh Tool and Main Menu>Preprocessor>Meshing>Size Cntrls.**

14.6.3.2 Controls Used for Mapped Meshing

You can specify that the program use all quadrilateral area elements, all triangle area elements, or all hexahedral (brick) volume elements to generate a mapped mesh. Mapped meshing requires that an area or volume be 'regular', that is, it must meet certain criteria.

For mapped meshing, element sizes are produced based on the current settings of **DESIZE**, along with **ESIZE**, **KESIZE**, **LESIZE** and **AESIZE** settings (**Main Menu > Preprocessor>Meshing>Size Cntrls>option**). SmartSizing [**SMRTSIZE**] cannot be used for mapped meshing.

14.6.4 Meshing Your Solid Model

Once you have built your solid model, established element attributes, and set meshing controls, you are ready to generate the finite element mesh. First, however, it is usually good practice to save your model before you initiate mesh generation:

Command (s): **SAVE**

GUI: Utility Menu>File>Save as Jobname.db

You may also want to turn on the "mesh accept/reject" prompt by picking **Main Menu>Preprocessor>Meshing>Mesher Opts.** This feature, which is available only through the **GUI**, allows you to easily discard an undesirable mesh.

If you are meshing multiple volumes or areas at one time, you should consider using the meshing option **By Size** so the mesh is created in the smallest volume or area first. This helps ensure that your mesh is adequately dense in smaller volumes or areas and that the mesh is of a higher quality.

14.7 Defining Material Properties

Most element types require material properties. Depending on the application, material properties can be linear or nonlinear.

As with element types and real constants, each set of material properties has a material reference number. The table of material reference numbers versus material property sets is called the material table. Within one analysis, you may have multiple material property sets (to correspond with multiple materials used in the model). The program identifies each set with a unique reference number.

While defining the elements, you point to the appropriate material reference number using the **MAT** command.

14.8 Applying Loads and Obtaining the Solution

In this step, the **SOLUTION** processor defines the analysis type and analysis options, apply loads, specify load step options, and initiate the finite element solution. You can also apply loads via the **PREP7** preprocessor.

14. 8. 1 Specifying the Analysis Type and Analysis Options

Specify the analysis type based on the loading conditions and the response you wish to calculate. For example, if natural frequencies and mode shapes are to be calculated, you would choose a modal analysis. You can perform the following analysis types in the program: static (or steady-state), transient, harmonic, modal, spectrum, buckling, and substructuring.

Not all analysis types are valid for all disciplines. Modal analysis, for example, is not valid for a thermal model. The analysis guides in the documentation set describe the analysis types available for each discipline and the procedures to do those analyses.

Analysis options allow you to customize the analysis type. Typical analysis options are the method of solution, stress stiffening on or off, and Newton-Raphson options.

After you have defined the analysis type and analysis options, the next step is to apply loads. Some structural analysis types require other items to be defined first, such as master degrees of freedom and gap conditions. The Structural Analysis Guide describes these items where necessary.

14. 8. 2 Applying Loads

The word loads as used in the documentation includes boundary conditions (constraints, supports, or boundary field specifications) as well as other externally and internally applied loads. Loads are divided into these categories:

- DOF Constraints.
- Forces.
- Surface Loads.
- Body Loads.
- Inertia Loads.
- Coupled-field Loads.

You can apply most of these loads either on the solid model (keypoints, lines and areas) or the finite element model (nodes and elements).

Two important load-related terms you need to know are load step and substep. A load step is simply a configuration of loads for which you obtain a solution. In a structural analysis, for example, you may apply wind loads in one load step and gravity in a second load step. Load steps are also useful in dividing a transient load history curve into several segments.

Substeps are incremental steps taken within a load step. You use them mainly for accuracy and convergence purposes in transient and nonlinear analyses. Substeps are also known as time steps-steps taken over a period of time.

14.8.3 Specifying Load Step Options

Load step options are options that you can change from load step to load step, such as number of substeps, time at the end of a load step, and output controls. Depending on the type of analysis you are doing, load step options may or may not be required. The analysis procedures in the analysis guide manuals describe the appropriate load step options as necessary.

14.8.4 Initiating the Solution

To initiate solution calculations, use either of the following:

Command (s): **SOLVE**

GUI: **Main Menu>Solution>Solve>Current LS**

Main Menu>Solution>Solution method

When you issue this command, the program takes model and loading information from the database and calculates the results. Results are written to the results file and also to the database. The only difference is that only one set of results can reside in the database at one time, while you can write all sets of results (for all substeps) to the results file.

You can solve multiple load steps in a convenient manner:

Command (s): **LSSOLVE**

GUI: **Main Menu>Solution>Solve>From LS Files**

14.9 Reviewing the Results

After the solution has been calculated, use the postprocessors to review the results. Two postprocessors are available: **POST1** and **POST26.**

• Use **POST1**, the general postprocessor, to review results at one substep (time step) over the entire model or selected portion of the model. The command for entering **POST1** is /**POST1** (**Main Menu>General Postproc**), valid only at the Begin level. You can obtain contour displays, deformed shapes, and tabular listings to review and interpret the results of the analysis. **POST1** offers many other capabilities, including error estimation, load case combinations, calculations among results data, and path operations.

• Use **POST26**, the time-history postprocessor, to review results at specific points in the model over all time steps. The command for entering **POST26** is /**POST26** (**Main Menu>TimeHist Postpro**), valid only at the Begin level. You can obtain graph plots of results data versus time (or frequency) and tabular listings. Other **POST26** capabilities include arithmetic calculations and complex algebra.

14.10 Structural Introductory Tutorial

This is a simple, single load step, structural static analysis of the corner angle bracket shown below. The upper left-hand pin hole is constrained (welded) around its entire circumference, and a tapered pressure load is applied to the bottom of the lower right-hand pin hole. The objective of the problem is to demonstrate the typical ANSYS analysis procedure. The US Customary system of units is used.

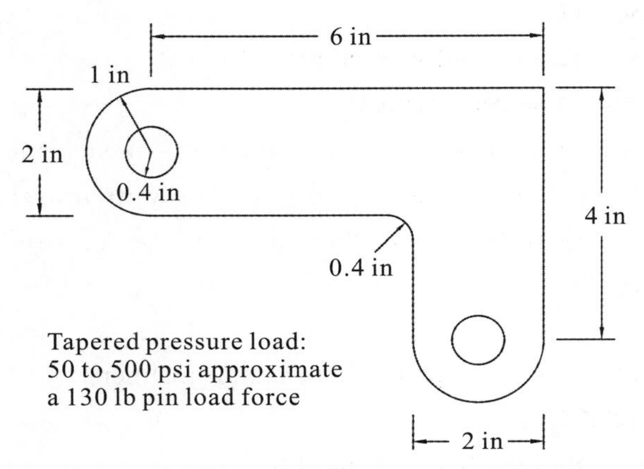

Fig. 14—11 Problem Description

14.10.1 Problem Given

The dimensions of the corner bracket are shown in the accompanying Fig. 14—11. The bracket is made of A36 steel with a Young's modulus of 3×10^7 psi and Poisson's ratio of 0.27.

14.10.2 Approach and Assumptions

Assume plane stress for this analysis. Since the bracket is thin in the z direction (1/2 inch thickness) compared to its x and y dimensions, and since the pressure load acts only in the x-y plane, this is a valid assumption.

Your approach is to use solid modeling to generate the 2-D model and automatically mesh it with nodes and elements. (Another alternative in ANSYS is to create the nodes and elements directly.)

14.10.3 Build Geometry

This is the beginning of Preprocessing.

Step 1: Define rectangles.

There are several ways to create the model geometry within ANSYS, some more convenient than others. The first step is to recognize that you can construct the bracket easily with combinations of rectangles and circle Primitives.

Decide where the origin will be located and then define the rectangle and circle primitives relative to that origin. The location of the origin is arbitrary. Here, use the center of the upper left-hand hole. ANSYS does not need to know where the origin is. Simply begin by defining a rectangle relative to that location. In ANSYS, this origin is called the **global origin.**

1. **Main Menu>Preprocessor>Modeling >Create>Areas>Rectangle>By Dimensions.**
2. Enter the following: (Note: Press the Tab key between entries)

   X1 =0 X2=6 Y1=−1 Y2=1
3. Apply to create the first rectangle.
4. Enter the following:

   X1 =4 X2=6 Y1=−1 Y2=−3
5. OK to create the second rectangle and close the dialog box.

Create Rectangle by Dimensions
[RECTNG] Create Rectangle by Dimensions
X1,X2 X-coordinates 2 0 6
Y1,Y2 Y-coordinates -1 1
OK 3 Apply Cancel Help

Create Rectangle by Dimensions
[RECTNG] Create Rectangle by Dimensions
X1,X2 X-coordinates 4 4 6
Y1,Y2 Y-coordinates -1 -3
5 OK Apply Cancel Help

Step 2: Change plot controls and replot.

The area plot shows both rectangles, which are areas, in the same color. To more clearly distinguish between areas, turn on area numbers and colors. The 'Plot Numbering Controls' dialog box on the Utility Menu controls how items are displayed in the Graphics Window. By default, a 'replot' is automatically performed upon execution of the dialog box. The replot operation will repeat the last plotting operation that occurred (in this case, an area plot).

1. **Utility Menu>Plot Ctrls>Numbering.**
2. Turn on area numbers.

3. OK to change controls, close the dialog box, and replot.

Plot Numbering Controls
[/PNUM] Plot Numbering Controls
KP Keypoint numbers — Off
LINE Line numbers — Off
AREA Area numbers — 2 On
VOLU Volume numbers — Off
NODE Node numbers — Off
Elem / Attrib numbering — No numbering
TABN Table Names — Off
SVAL Numeric contour values — Off
[/NUM] Numbering shown with — Colors & numbers
[/REPLOT] Replot upon OK/Apply? — Replot
3 OK — Apply — Cancel — Help

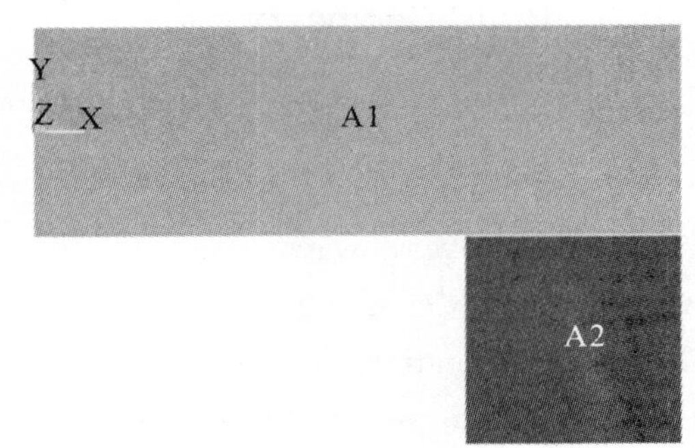

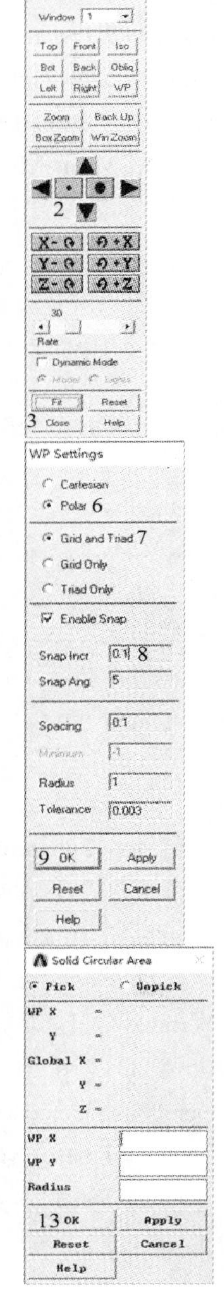

Before going to the next step, save the work you have done so far. ANSYS stores any input data in memory to the ANSYS database. To save that database to a file, use the SAVE operation, available as a tool on the Toolbar. ANSYS names the database file using the format jobname. db. If you started ANSYS using the product launcher, you can specify a jobname at that point (the default jobname is file).

You can check the current jobname at any time by choosing **Utility Menu>List>Status>Global Status.** You can also save the database at specific milestone points in the analysis (such as after the model is complete, or after the model is meshed) by choosing **Utility Menu>File>Save As** and specifying different jobnames (model. db, or mesh. db, and so on).

It is important to do an occasional save so that if you make a mistake, you can restore the model from the last saved state. You restore the model using the RESUME operation, also available on the Toolbar. (You can also find SAVE and RESUME on the Utility Menu, under File.)

4. Toolbar: SAVE_DB.

Step 3: Change working plane to polar and create first circle.

The next step in the model construction is to create the half circle at each end of the bracket. You will actually create a full circle on each end and then combine the circles and rectangles with a Boolean 'add' operation (discussed in step 5). To create the circles, you will use and display the working plane. You could have shown the working plane as you created the rectangles but it was not necessary.

Before you begin however, first 'zoom out' within the Graphics Window so you can see more of the circles as you create them. You do this using the 'Pan-Zoom-Rotate' dialog box, a convenient graphics control box you'll use often in any ANSYS session.

1. **Utility Menu>PlotCtrls>Pan, Zoom, Rotate.**
2. Click on small dot once to zoom out.
3. Close dialog box.
4. **Utility Menu>WorkPlane>Display Working Plane** (toggle on).

Notice the working plane origin is immediately plotted in the Graphics Window. It is indicated by the WX and WY symbols; right now coincident with the global origin X and Y symbols. Next you will change the WP type to polar, change the snap increment, and display the grid.

5. **Utility Menu>WorkPlane>WP Settings.**
6. Click on Polar.
7. Click on Grid and Triad.
8. Enter 0.1 for snap increment.
9. OK to define settings and close the dialog box.

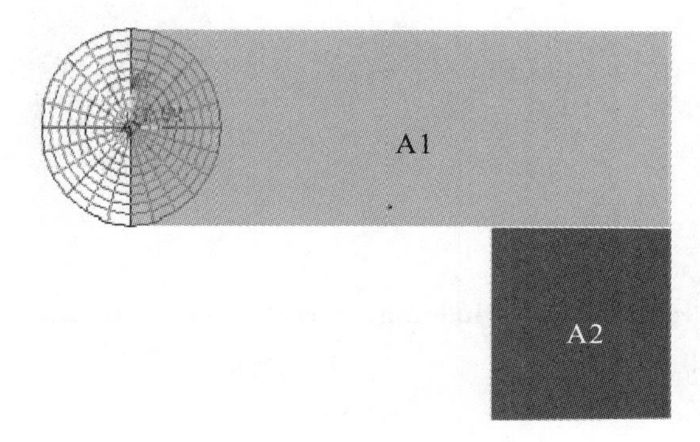

10. **Main Menu>Preprocessor>Modeling >Create >Areas>Circle>Solid Circle.**

Be sure to read prompt before picking.

11. Pick center point at:

WP X=0 (in Graphics Window shown below)

WP Y=0

12. Move mouse to radius of 1 and click left button to create circle.

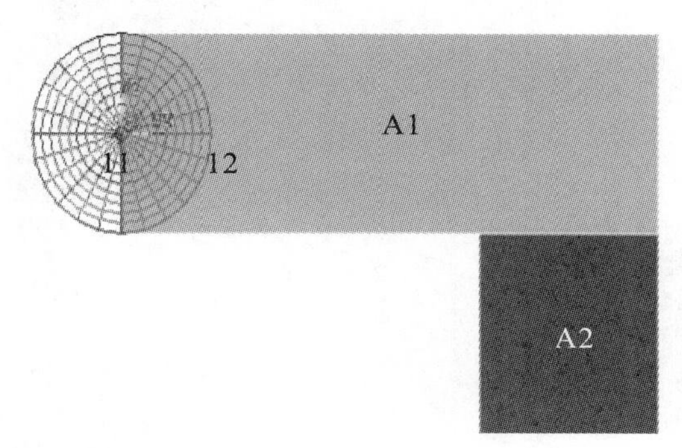

13. OK to close picking menu.
14. Toolbar: **SAVE _ DB.**

Step 4: Move working plane and create second circle.

To create the circle at the other end of the bracket in the same manner, you need to first move the working plane to the origin of the circle. The simplest way to do this without entering number offsets is to move the WP to an average keypoint location by picking the keypoints at the bottom corners of the lower, right rectangle.

1. **Utility Menu>WorkPlane>Offset WP to>Keypoints.**
2. Pick keypoint at lower left corner of rectangle.
3. Pick keypoint at lower right of rectangle.
4. OK to close picking menu.

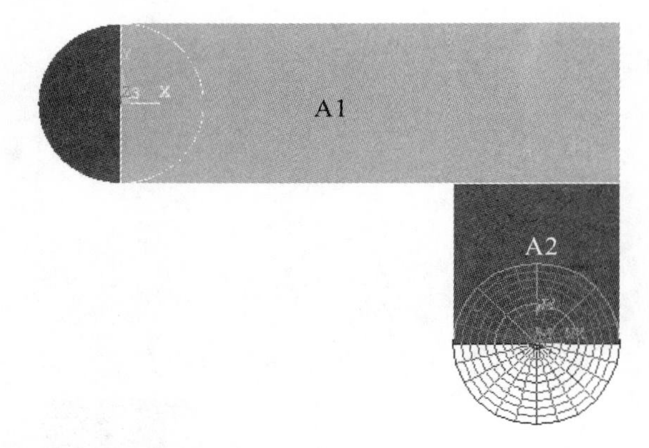

5. **Main Menu>Preprocessor>Modeling>Create>Areas>Circle>Solid Circle.**
6. Pick center point at:

   WP X=0

   WP Y=0
7. Move mouse to radius of 1 and click left button to create circle.

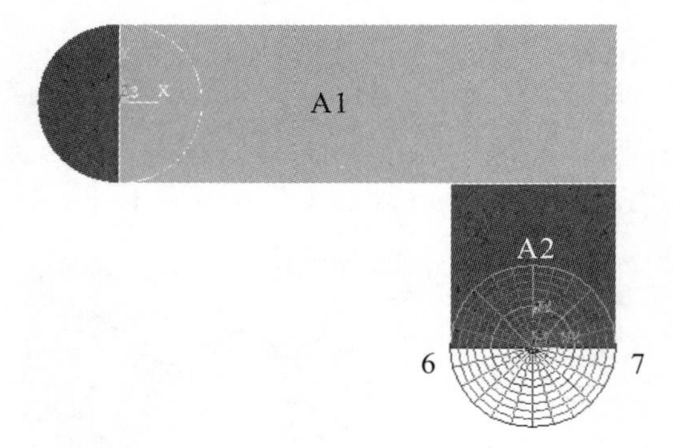

8. OK to close picking menu.
9. Toolbar: SAVE _ DB.

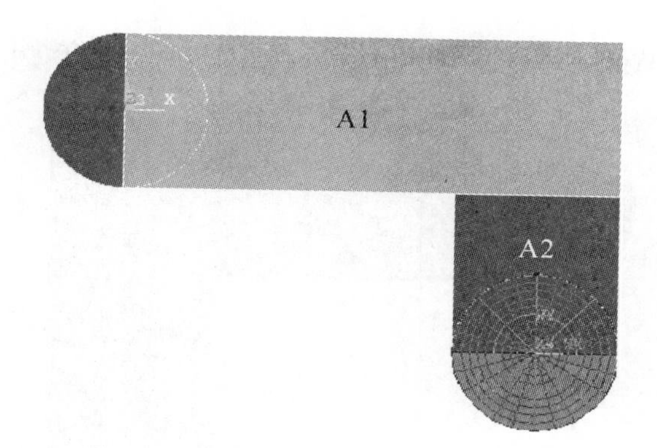

Step 5: Add areas.

Now that the appropriate pieces of the model are defined (rectangles and circles), you need to add them together so the model becomes one continuous piece. You do this with the Boolean add operation for areas.

1. **Main Menu>Preprocessor>Modeling>Operate>Booleans>Add>Areas.**
2. Pick All for all areas to be added.
3. Toolbar: **SAVE _ DB.**

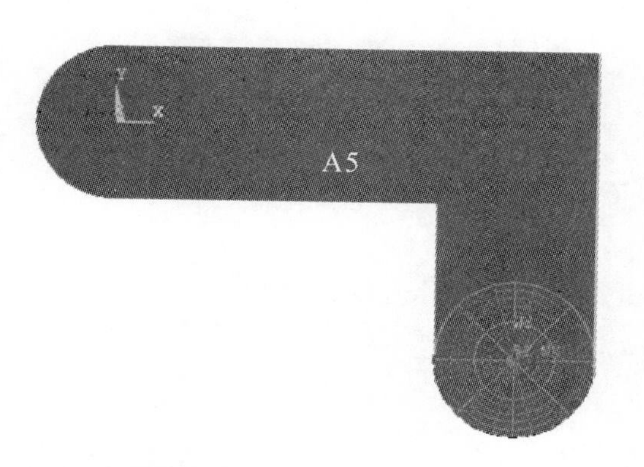

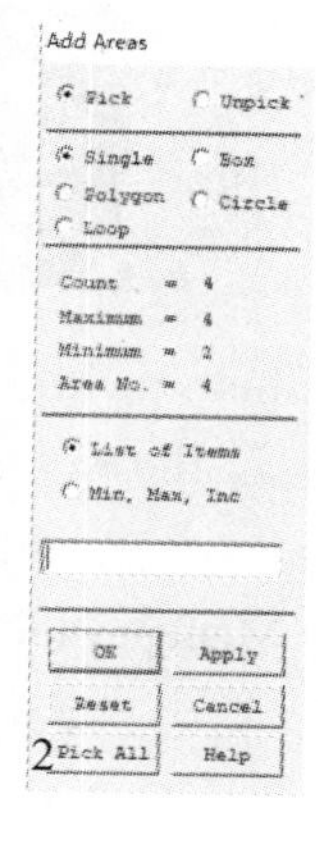

Step 6: Create line fillet.

1. **Utility Menu>PlotCtrls>Numbering.**
2. Turn on line numbering.
3. OK to change controls, close the dialog box, and automatically replot.

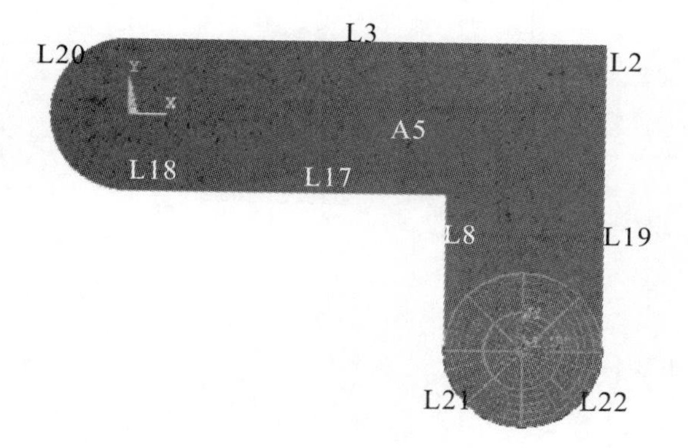

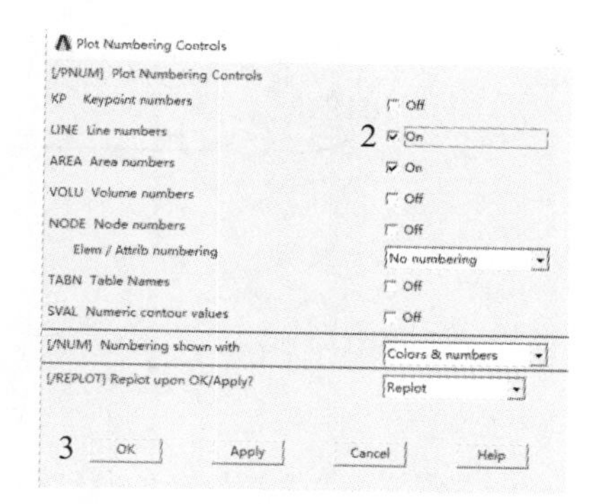

4. **Utility Menu>WorkPlane>Display Working Plane** (toggle off).

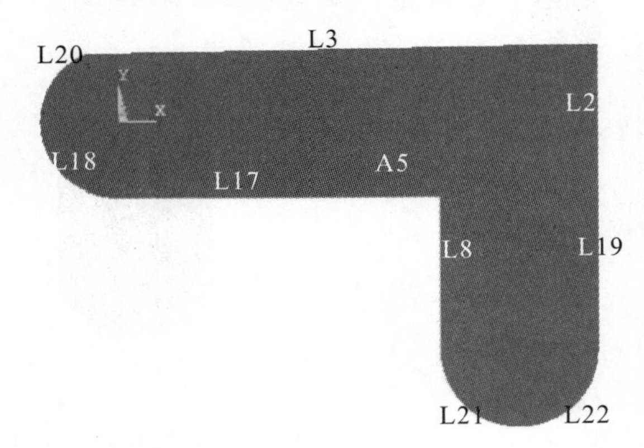

5. **Main Menu>Preprocessor>Modeling>Create>Lines>Line Fillet.**
6. Pick lines 17 and 8.

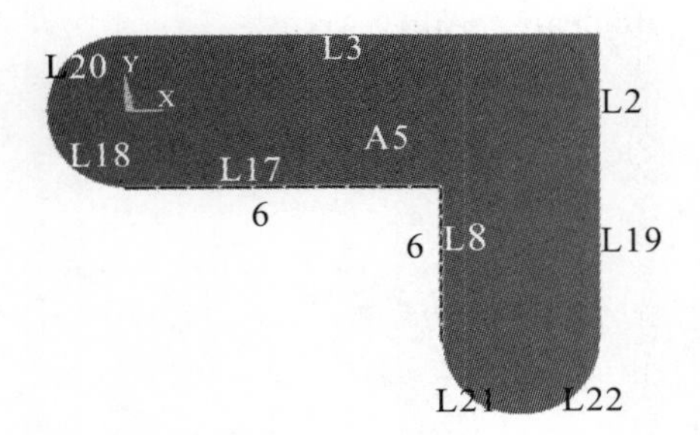

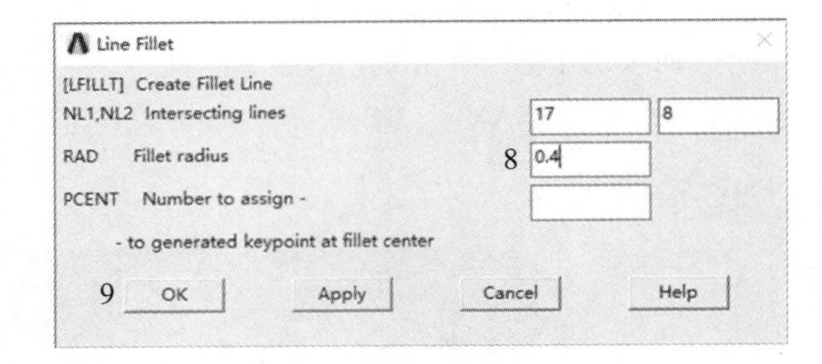

7. OK to finish picking lines (in picking menu).
8. Enter. 4 as the radius.
9. OK to create line fillet and close the dialog box.
10. **Utility Menu>Plot>Lines.**

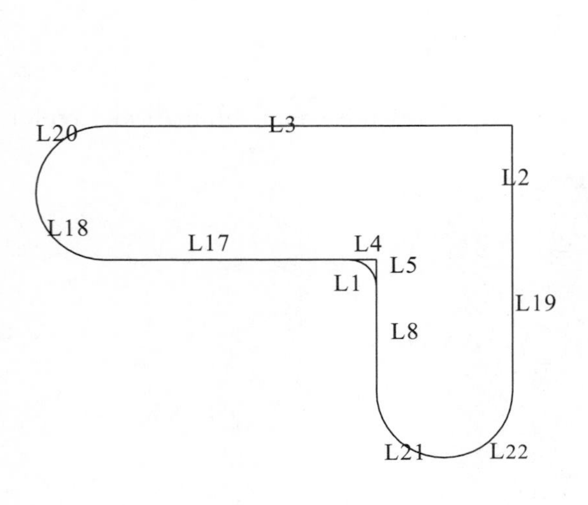

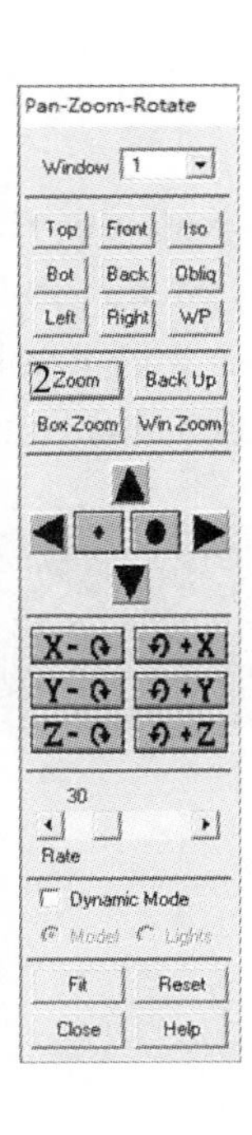

Step 7: Create fillet area.

1. **Utility Menu>PlotCtrls>Pan, Zoom, Rotate.**
2. Click on **Zoom** button.
3. Move mouse to fillet region, click left button, move mouse out and click again.

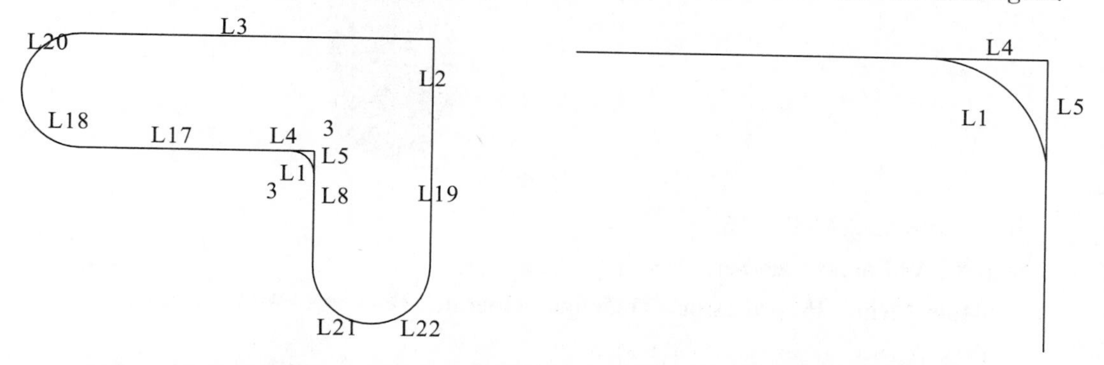

4. **Main Menu>Preprocessor>Modeling>Create>Areas>Arbitrary>By Lines.**

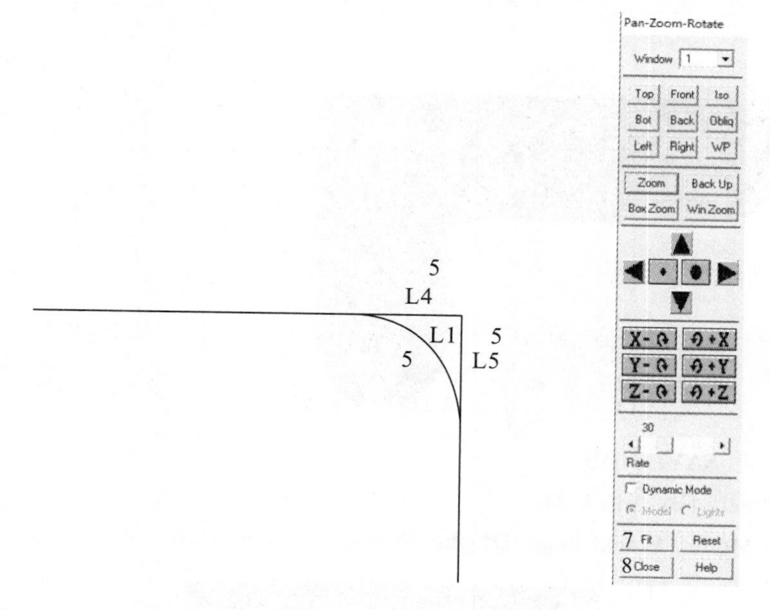

5. Pick lines 4, 5, and 1.
6. OK to create area and close the picking menu.
7. Click on Fit button.
8. Close the **Pan, Zoom, Rotate** dialog box.
9. **Utility Menu>Plot>Areas.**

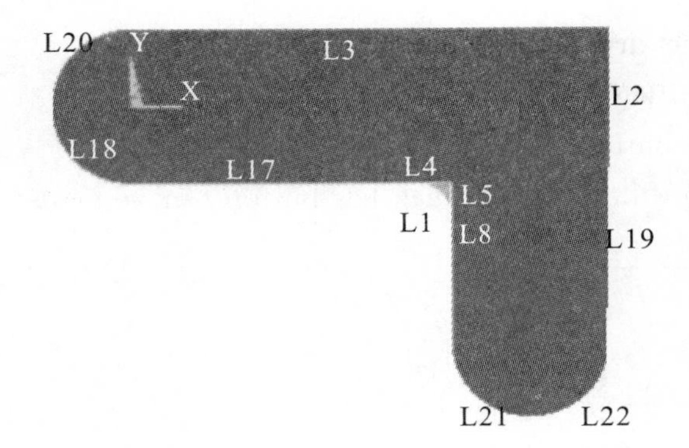

10. Toolbar: **SAVE _ DB.**

Step 8: Add areas together.

1. **Main Menu>Preprocessor>Modeling>Operate>Booleans>Add >Areas.**
2. Pick All for all areas to be added.

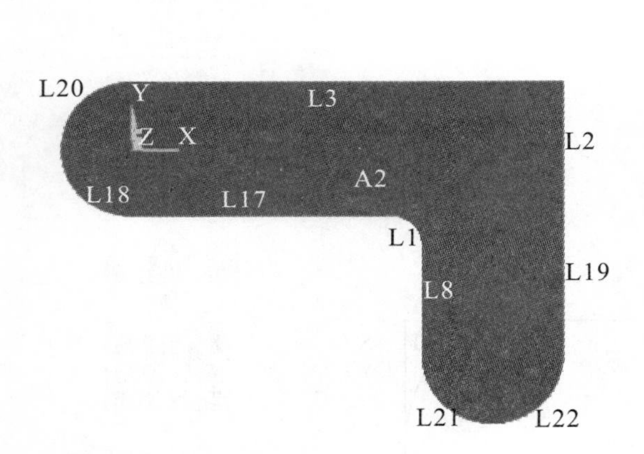

Add Areas

Pick Unpick

Single Box
Polygon Circle
Loop

Count = 0
Maximum = 2
Minimum = 2
Area No. =

List of Items
Min, Max, Inc

OK Apply
Reset Cancel
2 Pick All Help

3. Toolbar: **SAVE _ DB.**

Step 9: Create first pin hole.

1. **Utility Menu>WorkPlane>Display Working Plane** (toggle on).

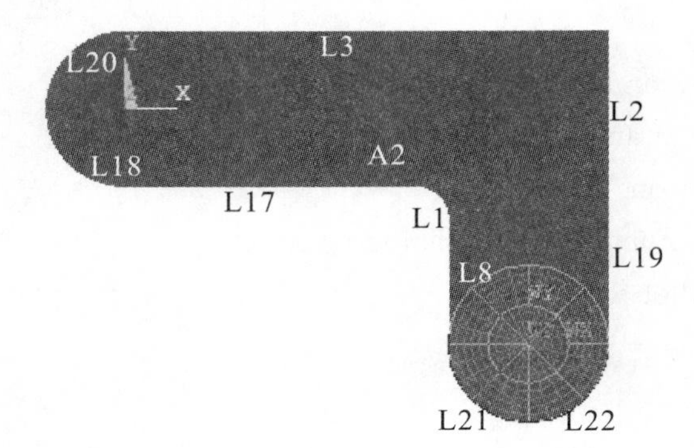

2. **Main Menu>Preprocessor>Modeling>Create>Areas>Circle>Solid Circle.**

3. Pick center point at:

WP X=0 (in Graphics Window)

WP Y=0

4. Move mouse to radius of 4 (shown in the picking menu) and click left button to create circle.

5. OK to close picking menu.

Step 10: Move working plane and create second pin hole.

1. **Utility Menu>WorkPlane>Offset WP to>Global Origin.**

2. **Main Menu>Preprocessor>Modeling>Create>Areas>Circle>Solid Circle.**

3. Pick center point at:

WP X=0 (in Graphics Window)

WP Y=0

4. Move mouse to radius of 4 (shown in the picking menu) and click left mouse button to create circle.

5. OK to close picking menu.

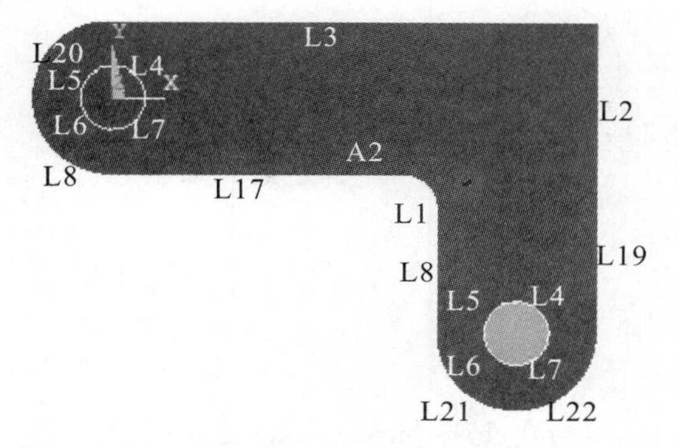

6. **Utility Menu>WorkPlane>Display Working Plane** (toggle off).

7. **Utility Menu>Plot>Replot.**

From this area plot, it appears that one of the pin hole areas is not there. However, it is there (as indicated by the presence of its lines), you just can't see it in the final display of the screen. That is because the bracket area is drawn on top of it. An easy way to see all areas is to plot the lines instead.

8. **Utility Menu>Plot>Lines.**

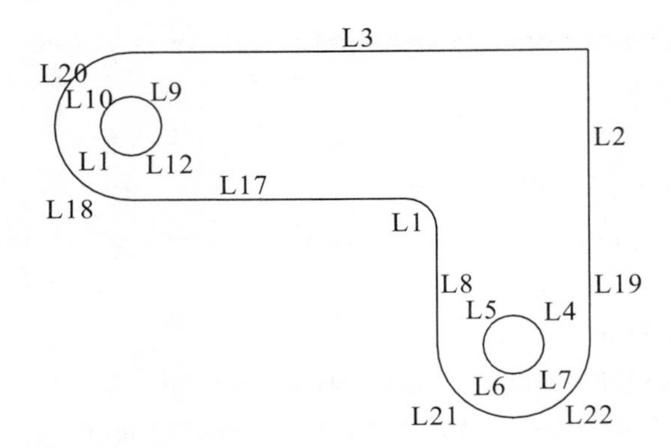

9. Toolbar: **SAVE _ DB.**

Step 11: Subtract pin holes from bracket.

1. **Main Menu>Preprocessor>Modeling>Operate>Booleans>Subtract>Areas.**
2. Pick bracket as base area from which to subtract.
3. Apply (in picking menu).
4. Pick both pin holes as areas to be subtracted.

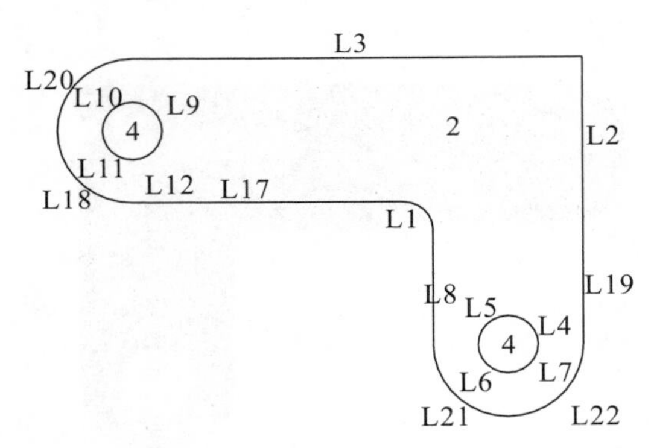

5. OK to subtract holes and close picking menu.

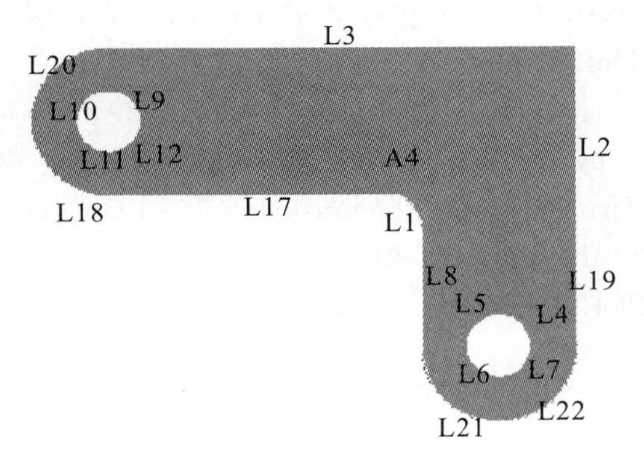

Step 12: Save the database asmodel. db.

At this point, you will save the database to a named file—a name that represents the model before meshing. If you decide to go back and remesh, you'll need to resume this database file. You will save it as model. db.

1. **Utility Menu>File>Save As.**
2. Enter model. db for the database file name.
3. OK to save and close dialog box.

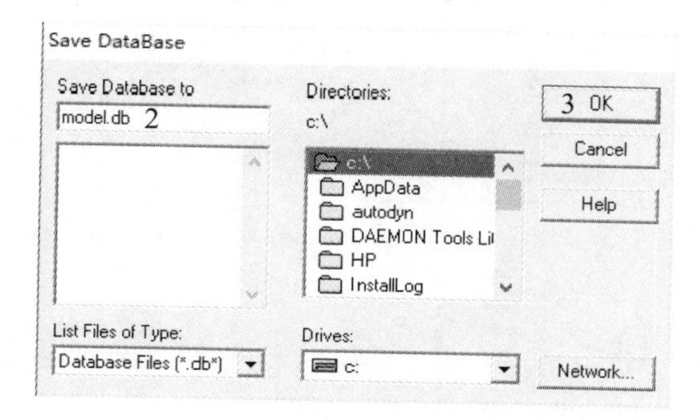

14. 10. 4 Define Materials

Step 13: Set preferences.

In preparation for defining materials, you will set preferences so that only materials that pertain to a structural analysis are available for you to choose.

To set preferences:

1. **Main Menu>Preferences.**
2. Turn on structural filtering. The options may differ from what is shown here since they depend on the ANSYS product you are using.
3. OK to apply filtering and close the dialog box.

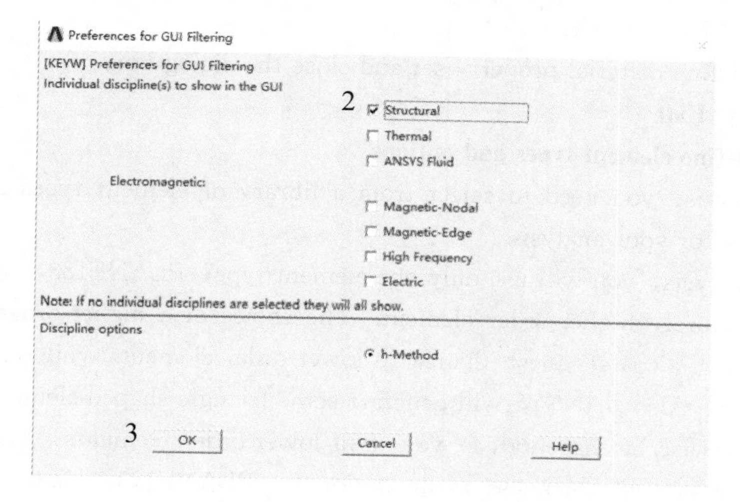

Step 14: Define material properties.

To define material properties for this analysis, there is only one material for the bracket, A36 Steel, with given values for Young's modulus of elasticity and Poisson's ratio.

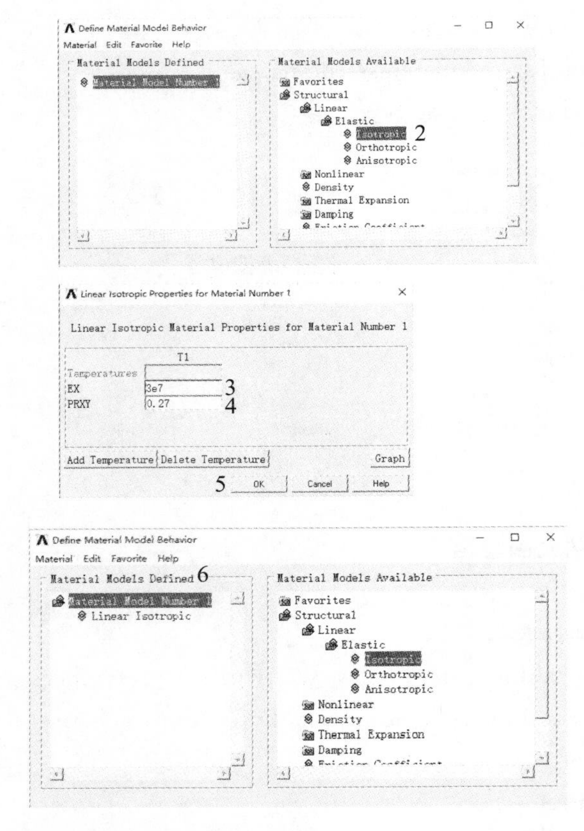

1. **Main Menu>Preprocessor>Material Props>Material Models.**
2. Double-click on Structural, Linear, Elastic, Isotropic.
3. Enter 30e6 for EX.
4. Enter 0.27 for PRXY.
5. OK to define material property set and close the dialog box.
6. **Material>Exit**

Step 15: Define element types and options.

In any analysis, you need to select from a library of element types and define the appropriate ones for your analysis.

For this analysis, you will use only one element type, PLANE183, which is a 2-D, quadratic, structural, higher-order element. The choice of a higher-order element here allows you to have a coarser mesh than with lower-order elements while still maintaining solution accuracy. Also, ANSYS will generate some triangle shaped elements in the mesh that would otherwise be inaccurate if you used lower-order elements. You will need to specify plane stress with thickness as an option for PLANE183. (You will define the thickness as a real constant in the next step.)

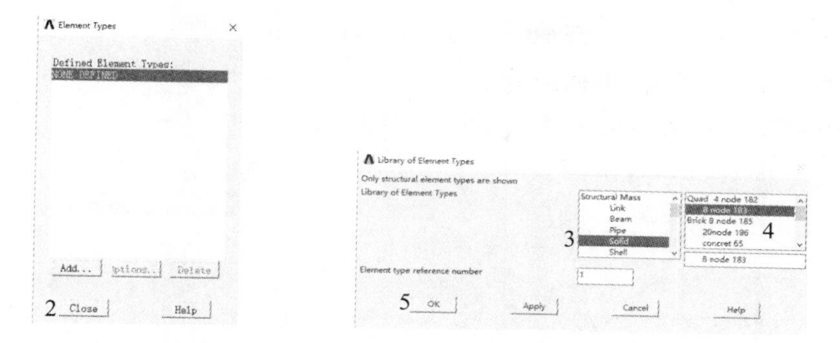

1. **Main Menu>Preprocessor>Element Type>Add/Edit/Delete.**
2. Add an element type.
3. Structural solid family of elements.
4. Choose the 8-node quad (PLANE183).
5. OK to apply the element type and close the dialog box.

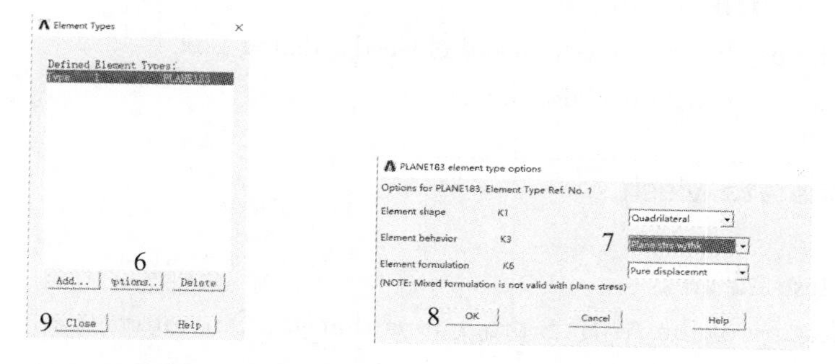

6. Options for PLANE183 are to be defined.
7. Choose plane stress with thickness option for element behavior.
8. OK to specify options and close the options dialog box.
9. Close the element type dialog box.

Step 16: Define real constants.

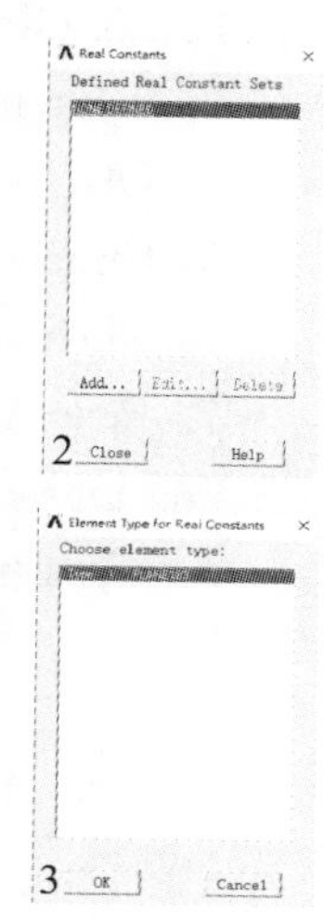

For this analysis, since the assumption is plane stress with thickness, you will enter the thickness as a real constant for PLANE183. To find out more information about PLANE183, you will use the ANSYS Help System in this step by clicking on a Help button from within a dialog box.

1. **Main Menu>Preprocessor>Real Constants>Add/Edit/Delete.**
2. Add a real constant set.
3. OK for PLANE183.

Before clicking on the Help button in the next step, you should be aware that the help information may appear in the same window as this tutorial, replacing the contents of the tutorial. After reading the help information, click on the Back button to return to this tutorial. If the help information appears in a separate window from the tutorial, minimize or close the help window after you read the help information.

4. Help to get help on **PLANE183.**
5. Hold left mouse button down to scroll through element description.

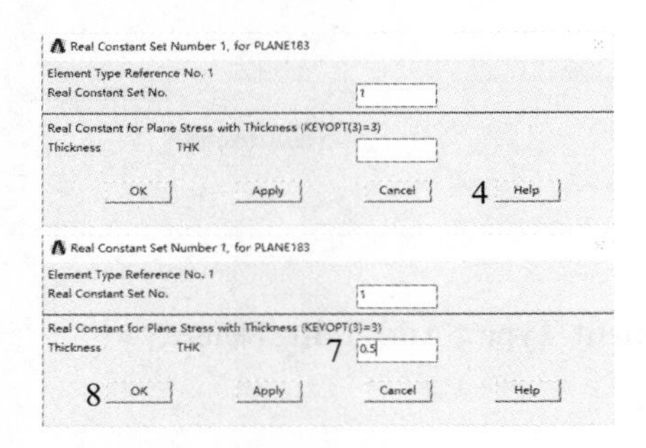

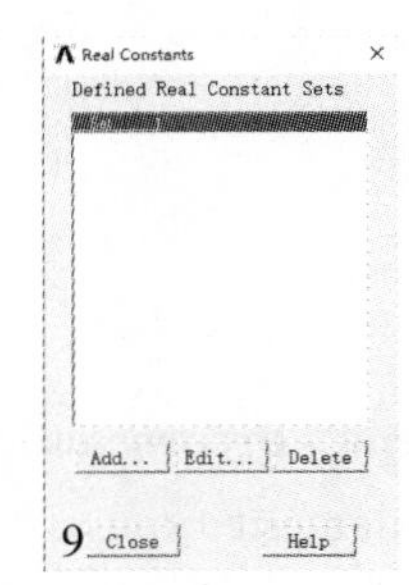

6. If the help information replaced the tutorial, click on the Back button to return to the tutorial.
7. Enter. 5 for **THK.**
8. OK to define the real constant and close the dialog box.
9. Close the real constant dialog box.

14. 10. 5 Generate Mesh

Step 17: Mesh the area.

One nice feature of the ANSYS program is that you can automatically mesh the model without specifying any mesh size controls. This is using what is called a default mesh. If you're not sure how to determine the mesh density, let ANSYS try it first! For this model, however, you will specify a global element size to control overall mesh density.

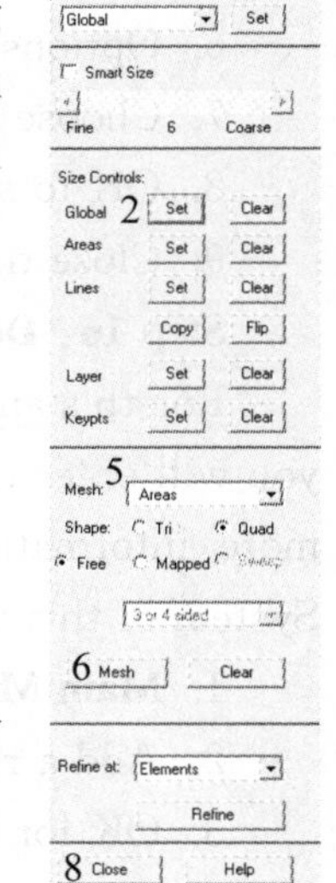

1. **Main Menu>Preprocessor>Meshing >Mesh Tool.**
2. Set Global Size control.
3. Type in 0. 5.
4. OK.
5. Choose Area Meshing.
6. Click on Mesh.
7. Pick All for the area to be meshed (in picking menu). Close any warning messages that appear.
8. Close the Mesh Tool.

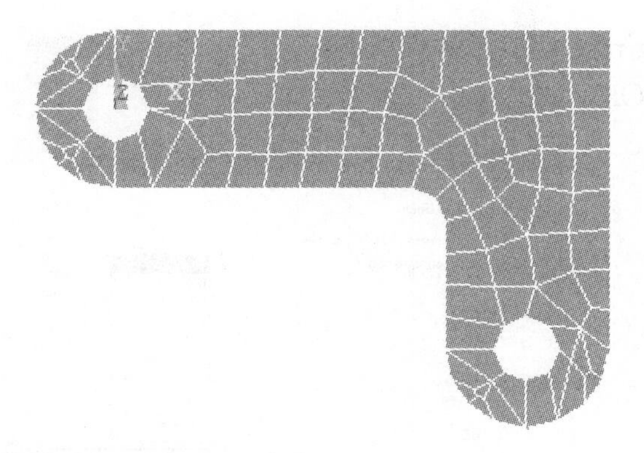

Step 18: Save the database as mesh. db.

Here again, you will save the database to a named file, this time mesh. db.

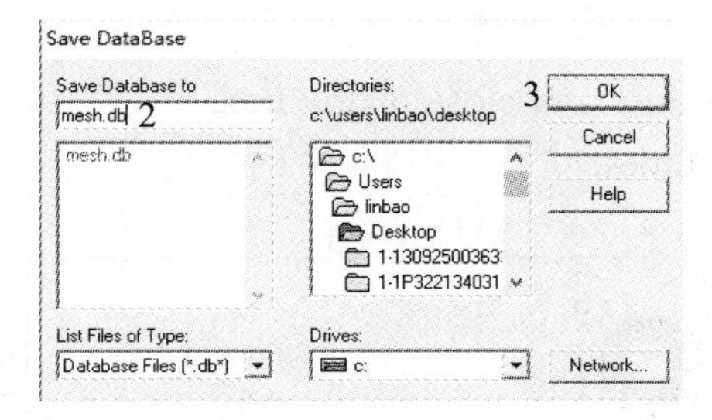

1. **Utility Menu>File>Save as.**
2. Enter mesh. db for database file name.
3. OK to save file and close dialog box.

14. 10. 6 Apply Loads

The beginning of the solution phase.

A new, static analysis is the default, so you will not need to specify analysis type for this problem. Also, there are no analysis options for this problem.

Step 19: Apply displacement constraints.

You can apply displacement constraints directly to lines.

1. **Main Menu>Solution>Define Loads >Apply>Structural>Displacement> On Lines.**
2. Pick the four lines around left-hand hole (Line numbers 10, 9, 11, 12).

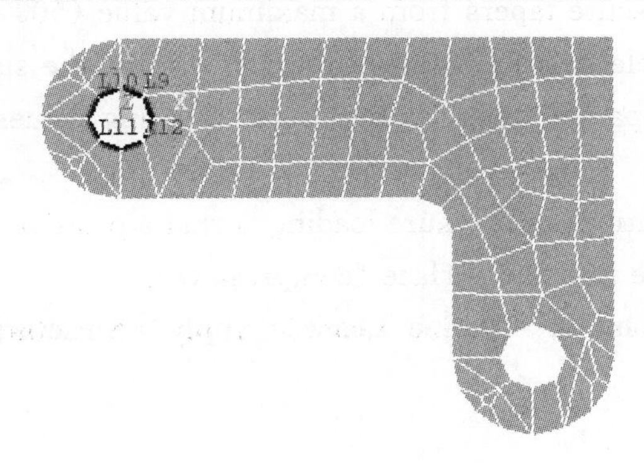

3. OK (in picking menu).
4. Click on All DOF.
5. Enter 0 for zero displacement.

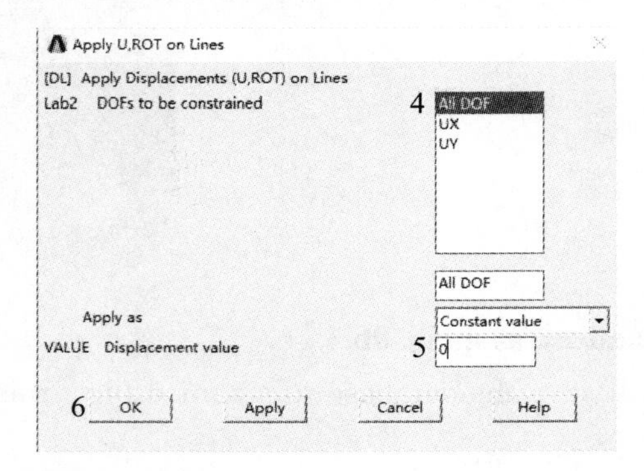

6. OK to apply constraints and close dialog box.
7. **Utility Menu>Plot Lines.**

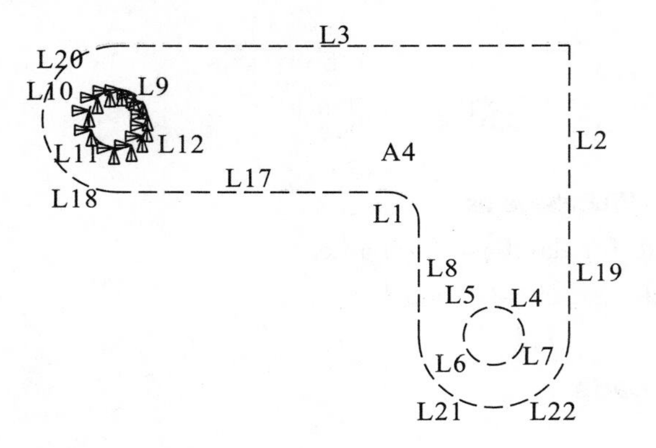

8. Toolbar: **SAVE _ DB.**

Step 20: Apply pressure load.

Now apply the tapered pressure load to the bottom, right-hand pin hole. ('Tapered' here means varying linearly.) Note that when a circle is created in ANSYS, four lines define the perimeter. Therefore, apply the pressure to two lines making up the lower half of the circle. Since the pressure tapers from a maximum value (500 psi) at the bottom of the circle to a minimum value (50 psi) at the sides, apply pressure in two separate steps, with reverse tapering values for each line.

The ANSYS convention for pressure loading is that a positive load value represents pressure into the surface (compressive).

1. **Main Menu > Solution > Define Loads > Apply > Structural > Pressure>On Lines.**

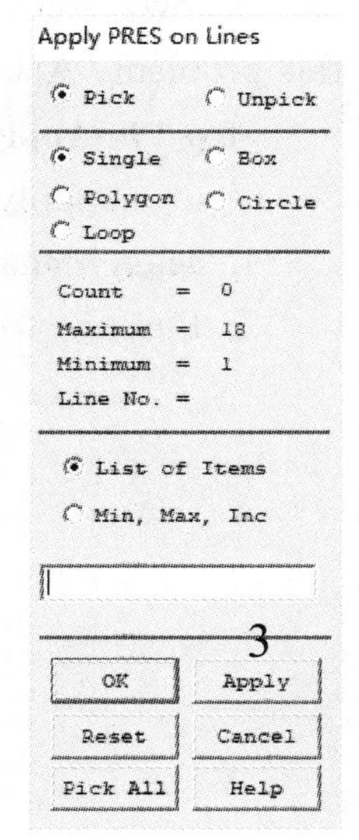

2. Pick line defining bottom left part of the circle (line 6).

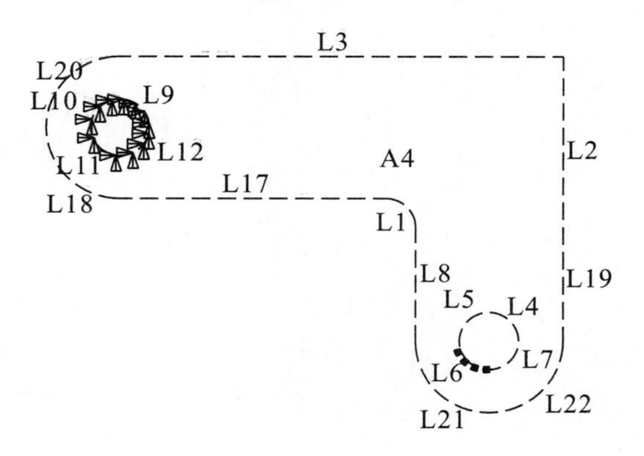

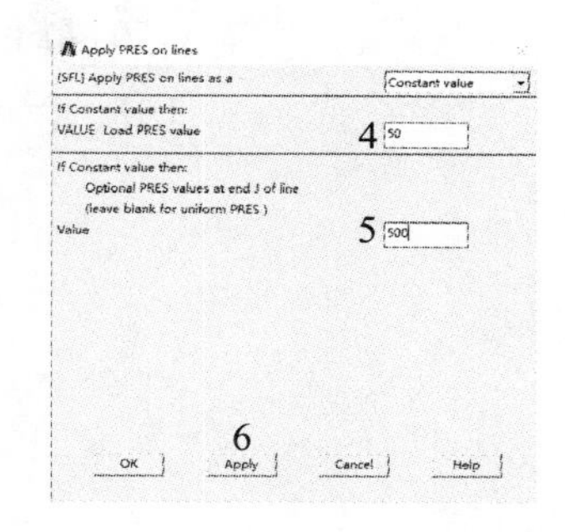

3. Apply.
4. Enter 50 for VALUE.
5. Enter 500 for optional value.
6. Apply.
7. Pick line defining bottom right part of circle (line 7).

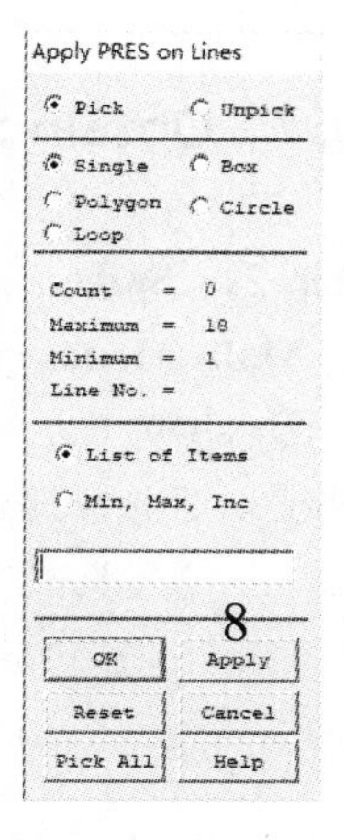

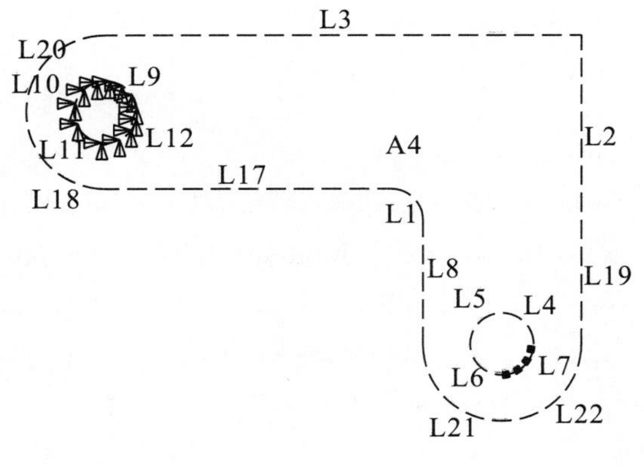

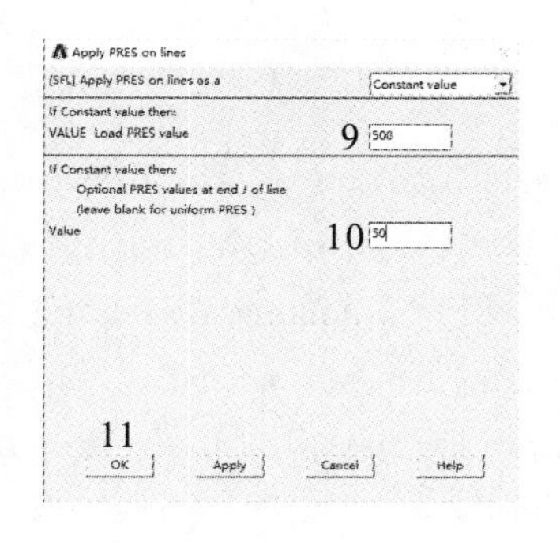

8. Apply.
9. Enter 500 for VALUE.
10. Enter 50 for optional value.

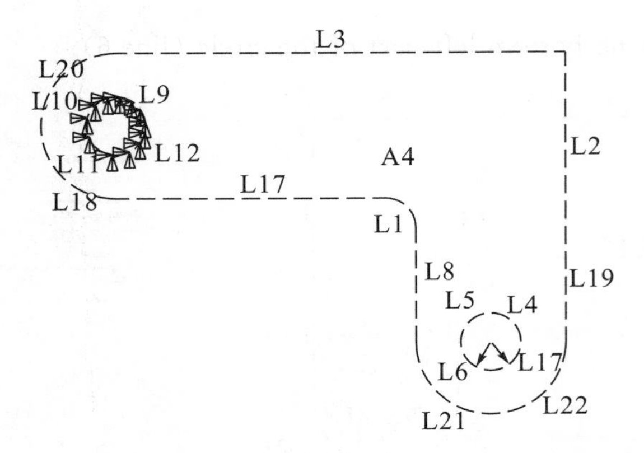

11. OK.

12. Toolbar: **SAVE _ DB.**

14.10.7 Obtain Solution

Step 21: Solve.

1. **Main Menu>Solution>Solve>Current LS.**

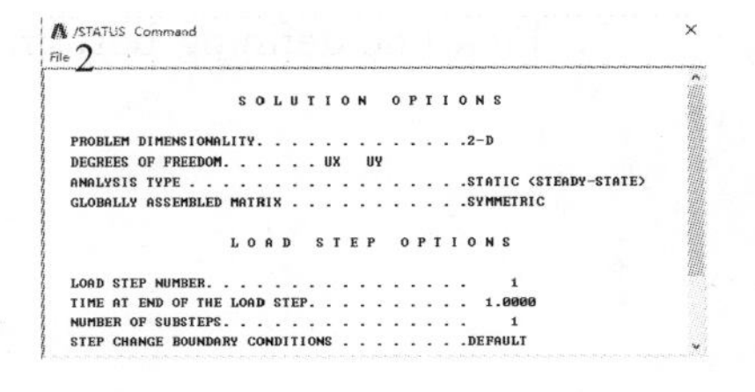

2. Review the information in the status window, then choose File>Close (Windows), or Close (Linux), to close the window.

3. OK to begin the solution. Choose Yes to any Verify messages that appear.

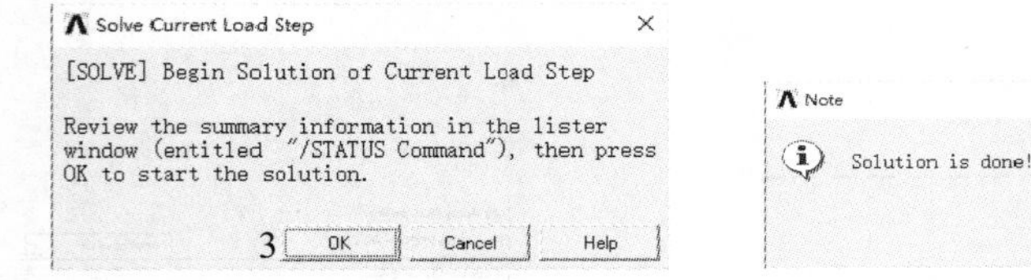

4. Close the information window when solution is done.

ANSYS stores the results of this one load step problem in the database and in the results file, **Jobname. RST** (or **Jobname. RTH** for thermal, **Jobname. RMG** for magnetic). The database can actually contain only one set of results at any given time, so in a multiple load step or multiple substep analysis, ANSYS stores only the final solution in the database. ANSYS stores all solutions in the results file.

14.10.8 Review Results

The beginning of the postprocessing phase.

Step 22: Enter the general postprocessor and read in the results.

1. **Main Menu>General Postproc>Read Results>First Set.**

Step 23: Plot the deformed shape.

1. **Main Menu>General Postproc>Plot Results>Deformed Shape.**
2. Choose Def+undeformed.

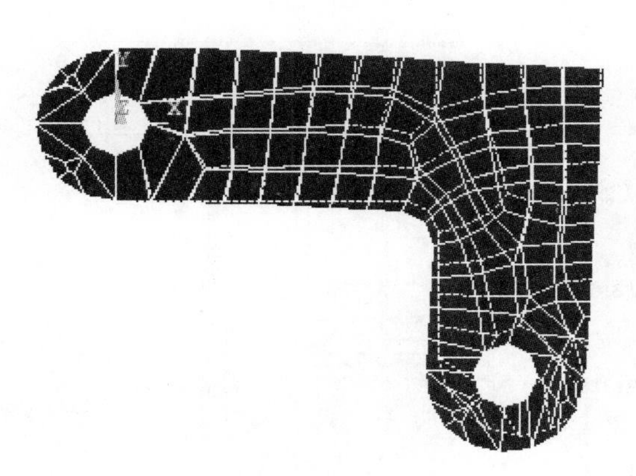

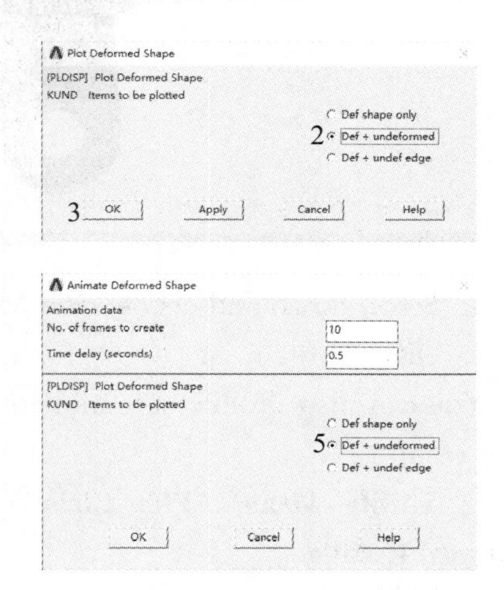

3. OK.

You can also produce an animated version of the deformed shape.

4. **Utility Menu>Plot Ctrls>Animate>Deformed Shape.**
5. Choose Def +undeformed.

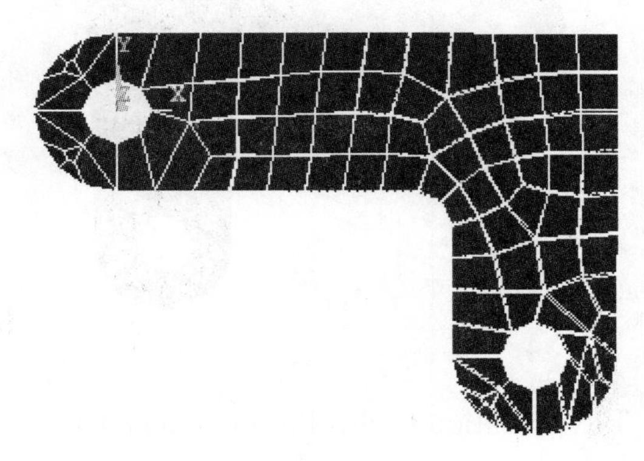

6. OK.
7. Make choices in the Animation Controller (not shown), if necessary, then choose **Close.**

Step 24: Plot the von Mises equivalent stress.

1. **Main Menu>General Postproc>Plot Results>Contour Plot>Nodal Solu.**
2. Choose Stress item to be contoured.

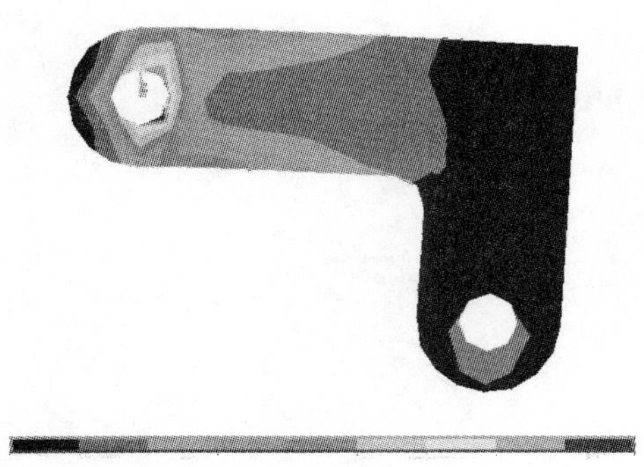

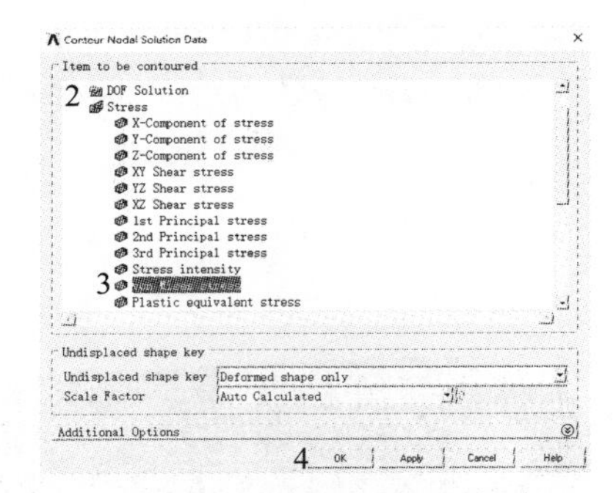

3. Scroll down and choose von Mises (SEQV).

4. OK.

You can also produce an animated version of these results.

5. **Utility Menu > Plot Ctrls > Animate > Deformed Results.**

6. Choose Stress item to be contoured.

7. Scroll down and choose von Mises (SEQV).

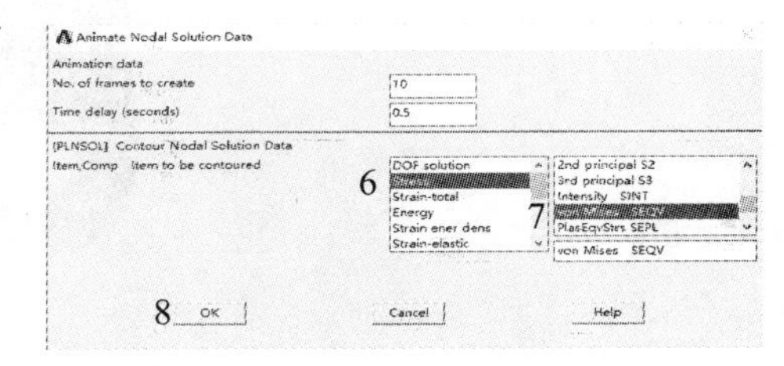

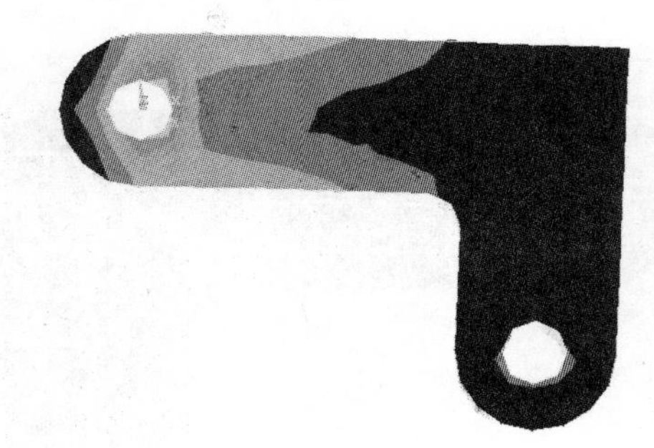

8. OK.

9. Make choices in the Animation Controller (not shown), if necessary, then choose **Close.**

Step 25: List reaction solution.

1. **Main Menu>General Postproc>List Results>Reaction Solu.**

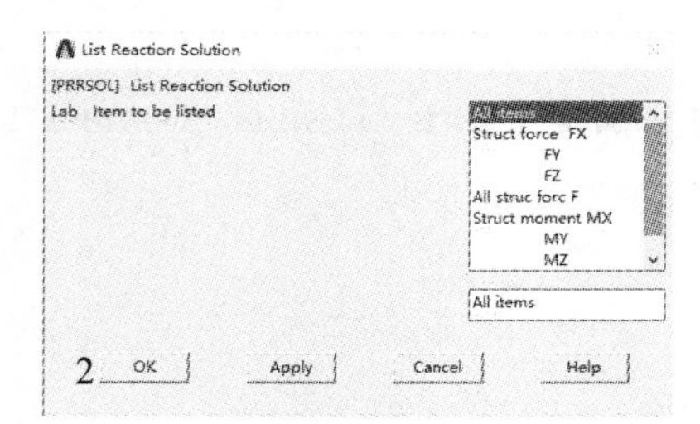

2. OK to list all items and close the dialog box.

3. Scroll down and find the total vertical force, FY.

PRRSOL Command
File 4

```
TIME=    1.0000       LOAD CASE=   0

THE FOLLOWING X,Y,Z SOLUTIONS ARE IN THE GLOBAL COORDINATE SYSTEM

  NODE       FX          FY
   117    139.18      -4.8320
   118   -23.436       101.83
   119    153.77       46.132
   120    105.93       41.964
   121    88.078       106.53
   122   -94.004       7.9702
   123   -23.085       104.59
   124   -140.52       45.532
   125   -203.23       35.963
   126   0.42687      -66.958
   127   -81.655      -51.404
   128   -43.265      -30.963
   129   -52.291      -62.370
   130    50.855      -61.913
   131    40.625      -27.897
   132    82.629      -49.568

TOTAL VALUES
VALUE  -0.31976E-07  134.61   3
```

4. **File>Close** (Windows), or Close (Linux), to close the window.

The value of 134.61 is comparable to the total pin load force.

There are many other options available for reviewing results in the general postprocessor. You'll see some of these demonstrated in other tutorials. You have finished the analysis. Exit the program in the next step.

Step 26: Exit the ANSYS program.

When exiting the ANSYS program, you can save the geometry and loads portions of the database (default), save geometry, loads, and solution data (one set of results only), save geometry, loads, solution data, and postprocessing data (i.e., save everything), or save nothing. You can save nothing here, but you should be sure to use one of the other save options if you want to keep the ANSYS data files.

1. Toolbar: **Quit.**
2. Choose **Quit**-No Save!

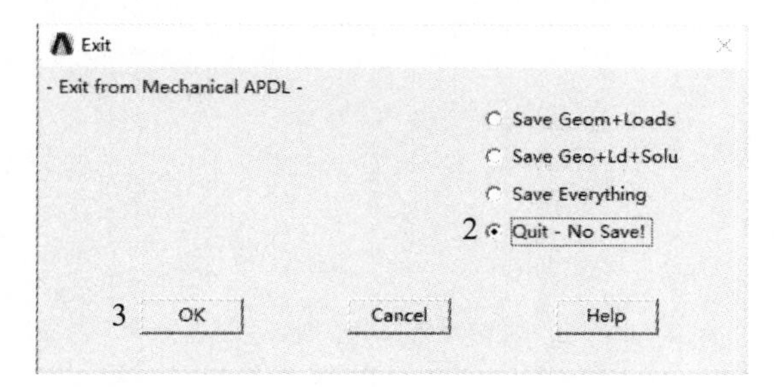

3. OK.

Congratulations! You have completed this tutorial.

Even though you have exited the ANSYS program, you can still view animations using the ANSYS ANIMATE program. The ANIMATE program runs only on the PC and is extremely useful for:

- Viewing ANSYS animations on a PC regardless of whether the files were created on a PC (AVI files) or on a Linux workstation (ANIM files).
- Converting ANIM files to AVI files.
- Sending animations over the web.